Unequal City

Unequal City examines some of the dramatic economic and social changes that have taken place in London over the last forty years. It describes how London's changing industrial structure, particularly the shift from an industrial- to a services-based city, and the associated changes in occupational class structure and in the structure of earnings and incomes, have worked through to the housing market and the gentrification of large parts in Inner London. This has had major consequences for both the social structure and the built environment of London.

It asserts that this transformation in London's industrial structure, from a city with a large manufacturing base to one based primarily on business, financial, creative and other services, is linked to a major change in its occupational structure. But, unlike much of the literature, which argues that cities like London have become increasingly socially polarised, it argues that London has become more professionalised and has a shrinking manual workforce. The changes in the industrial and occupational structure of London have been linked to changes in its earnings and income structures. The dominant feature of London has been an expansion of its high-earning groups and a marked increase in both earnings and income inequality. The growth of this expanded new middle class has had major impacts on the nature of the London housing market, particularly in the growth of home ownership, rising prices and the expansion of middle-class gentrification across much of Inner London. This has been paralleled by the growing marginalisation of the less skilled, the unemployed and various minority groups in the council sector. These changes have reshaped the social structure and social geography of London. They have made it a more unequal city.

Unequal City relates to the literature on global cities. The book has a wide sweep and summarises a wide range of literature on occupational and industrial change, earnings and incomes and the housing market and gentrification. It provides a wealth of original data, figures, maps and tables and will be a valuable reference for anyone interested in the changes that have reshaped the social structure of London in recent decades.

Chris Hamnett is Professor of Human Geography at King's College London.

Unequal City

London in the global arena

Chris Hamnett

Routledge
Taylor & Francis Group

LONDON AND NEW YORK

First published 2003
by Routledge
2 Park Square, Milton Park, Abingdon, Oxon, OX14 4RN

Simultaneously published in the USA and Canada
by Routledge
270 Madison Ave, New York NY 10016

Reprinted 2004

Routledge is an imprint of the Taylor & Francis Group

Transferred to Digital Printing 2005

© 2003 Chris Hamnett

Typeset in 10/12pt Times NR by Graphicraft Limited, Hong Kong

British Library Cataloguing in Publication Data
A catalogue record for this book is available from the
British Library,

Library of Congress Cataloging in Publication Data
A catalog record for this book has been requested

ISBN 0 415 31730 4 (hbk)
ISBN 0 415 31731 2 (pbk)

Printed and bound by Antony Rowe Ltd, Eastbourne

Contents

Plates

Figures

Tables

Acknowledgements

The author and publishers would like to thank the following for granting permission to reproduce material in this book:

The City Fringe Partnership for Figure 9.1.
Corporation of London for Table 2.8.
The Greater London Authority for Figures 6.8, 7.2, 8.1–8.6 and Table 5.4.
Hutchison Wampoa Property/Chestertons for Plate 9.2.
Income Data Services for Table 4.10.
Lambeth Borough Council for Figure 9.3.
The Land Registry for Figure 6.6 and Tables 6.3, 6.4.
London Stock Exchange for Table 2.5.
Office for National Statistics for data in Figures 2.1, 3.1–3.3, 4.1, 6.1, 6.7, 7.3–7.5 and in Tables 2.1–2.4, 2.6, 3.2, 3.3, 4.1–4.3, 4.5–4.9, 4.11–4.16, 5.1–5.3, 6.1, 6.2, 8.1–8.3.
Pion Ltd for Figures 4.1a, 4.1b, 6.5a, 6.5b, 6.5c, 9.4a, 9.4b, 9.4c and Table 4.5.
Taylor & Francis/ *Urban Studies* (www.tandf.co.uk) for Figures 5.5, 5.8 and Table 7.1.
Taylor & Francis/*Journal of Ethnic and Migration Studies* (www.tandf.co.uk) for Figure 5.7.

Chapter 1

Introduction

The remaking of London

This book is concerned with the transformation in the economic, social and physical structure of London which has taken place over the last 30–40 years. It examines in detail the nature of this transformation (or rather the set of linked transformations), its scale and extent, causes and consequences, and puts forward an explanation focused on the changing economy of London and its global role, which tries to account for it. There can be little doubt that London has changed dramatically over this period. If we start with a summary overview back to the early 1960s, London was still a city with a large manufacturing base and a large working-class population. The physical structure of Inner London was dominated by Victorian terraced houses and its housing market was still overwhelmingly privately rented with few council blocks and limited owner-occupation. The docks were a vital part of the economy of the East End, along with the furniture and printing industries, the City of London financial area was relatively small scale, Covent Garden was still London's fruit and vegetable market, much of the riverside was run down and Canary Wharf did not exist. Ethnic minorities were small in scale and extent, and outside the traditional middle-class residential areas of South Kensington, St John's Wood and Hampstead, most of Inner London was still the preserve of the traditional Labour voting working class, most of whom lived in poor-quality terraced rented housing. Notting Hill was a run-down residential area with a small West Indian population, and Aldgate and Whitechapel, immediately to the east of the City of London, still had a relatively large Jewish population. The ring of districts surrounding the City – Clerkenwell, Shoreditch, Hoxton and Bethnal Green – were dominated by light industry and resolutely working class. Bermondsey, south of the river, was still a tough docklands community, as was the Isle of Dogs and much of Tower Hamlets. Islington, Camden Town and Kentish Town were run down inner-city districts, and upper Holloway had some of the worst housing conditions in London (see Figure 1.1). A little of the flavour of this rundown London is captured in the early 1980s film *The Long Good Friday*, with Bob Hoskins as the traditional, but completely out of his depth, East End gang boss trying to move into the new world of international finance and property development.

This traditional London began to change dramatically in the 1960s and early 1970s as manufacturing industry declined, as new council estates replaced poor-quality private renting and as the occupational structure of London began to change. In retrospect, and according to *Time Magazine* (1969), the advent of the 'Swinging Sixties' marked the start of a transformation from the old to the new London. Today, the face of London is very different from that of even thirty years ago. The Docks and old warehouses have been replaced by an imposing financial centre in Canary Wharf and by luxury waterfront apartments. Manufacturing industry has all but disappeared, as has manufacturing employment, and the vast majority of London's workforce now works in the service sector with a large proportion in financial and business services. Covent Garden is now one of London's fashionable shopping areas. Notting Hill and Islington are among London's most desirable and expensive residential areas, and much of those parts of Inner London which do not consist of high-density and high-rise council estates have either been gentrified or are in the process of being so. The City Fringe districts of Clerkenwell, Hoxton and Shoreditch are being converted into luxury loft apartments and Brick Lane, Whitechapel and Bow are being gentrified. The old East End is gradually being turned into expensive residential districts for City workers and the traditional working class is in retreat. Ethnic minorities now account for a third of the population of Inner London and London as a whole is becoming increasingly 'multi-ethnic'. It is possible to still find the traditional white working-class London of yesteryear but, increasingly, it is necessary to look to some of the suburban areas like Barking and Dagenham or to some of the inner-city council estates. In the space of less than forty years London has been transformed from an industrial to a post-industrial city, with a large proportion of its labour force working in financial and business services. The London of *The Long Good Friday* has given way to the London depicted in the films *Notting Hill* and *Sliding Doors*. The scale of the transformation was captured by a *Newsweek* cover story in 1999, headlined 'London: coolest city on the planet'. Even allowing for some journalistic exaggeration, it is clear that there have been major changes in the social and cultural life of the city since the 1960s.

London has also changed politically. With the exception of middle-class areas such as Kensington and Westminster, most of the working-class Inner City used to be strongly Labour, and the political map of London consisted of a small Conservative core, a large Labour inner ring, and a Conservative suburban periphery (Johnston *et al.*, 1988). This pattern changed substantially during the 1980s and 1990s, with many once-Labour-dominated inner-city areas shifting to Conservative or Liberal Democrat control as the social composition of the electorate changed and the previously solidly working-class inner areas were gentrified and became more socially mixed. Although

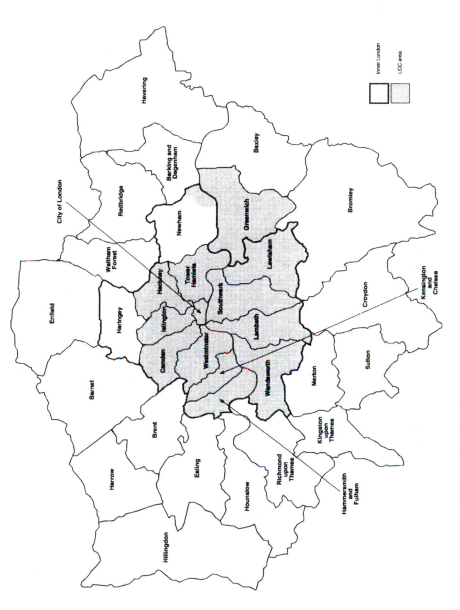

Inner London

LCC area

Figure 1.1 Map of Greater London, showing Inner and Outer boroughs and LCC.
Source: Chris Hamnett.

Labour's sweeping General Election victory in 1997 increased the number of London Labour MPs, the class composition of Inner London has irrevocably changed.

Last, but not least, the long-term population decline of London seems to have finally halted and reversed. From its peak of 8.6 million in 1939, Greater London's population fell sharply during the 1960s and 1970s, particularly in Inner London, as a combination of local authority redevelopment and gentrification led to the replacement of multiply-occupied overcrowded houses with single-family occupancy (Hamnett and Randolph, 1982). By 1981 Greater London's population had fallen to 6.8 million. But during the second half of the 1980s the population began to grow and the 2001 census suggests that it is now back to 7.2 million and is growing by 50,000 per year. If current trends continue it is projected to rise towards 8 million by 2020 (Greater London Authority, 2002). London has grown by the equivalent of a city the size of Frankfurt over the past fifteen years and could do so again in the next fifteen. Whether this will happen, or whether London's growth will be choked off by its growing housing shortage and increasing housing costs, remains to be seen (*Economist*, 2002). A great deal also hinges on the continued success of the City of London and its allied business service sector, now, arguably, the key motor of London's economy. If this seriously falters, the attraction of London as a place to live and work could take a serious dent.

Interpreting the changes

How are we to understand and interpret these changes. There are a number of possible approaches. First, it could be argued that the transformation which has taken place in London reflects its unique and specific history and bears little or no relation to changes in other cities. This argument is both theoretically and empirically untenable. Many of the changes which have taken place in London bear a close similarity to those in other major international cities, though they are not identical. Second, it can be argued, after Toulouse (1992), that the changes are primarily political in origin, a result of the rise of Thatcherism and the policies which were pursued by Conservative governments from 1979 to 1997. While there is some validity in this argument, it is argued that it is too limited, accords too much importance to politics and ignores the third explanation – namely, the fundamental importance of changes in the role and structure of London's economy in generating a set of related transformations in the structure of occupations, earnings, household incomes and the housing market. The argument of this book is based on three underlying propositions. First, that in order to understand the basis of these changes, we need to understand London's changing role in the global economy and financial system. During the 1970s, 1980s and 1990s London has strengthened its role as one of the major control centres for the global economic and financial system. It functions as one of the major 'world cities' (Friedmann, 1986) or 'global cities' as Sassen (1991) prefers, and exercises a leading role in the organisation and control of the world's economy, trade and financial flows. Second, it argues that the dramatic changes that have taken place in London over the last forty years are not isolated but are, instead, closely interrelated. The transformation in the industrial structure of London has been important in generating related changes in its occupational structure, its earnings and incomes structure, its housing market, social and ethnic composition, its geography, and its physical landscape. Third, it is argued that the changes are, to a significant extent, linked to the transformation of London from an *industrial* to a *post-industrial* city, whose economy is no longer dominated by manufacturing industry but is, instead, based on finance, business services and the creative and cultural industries (Greater London Authority, 2002).

London in the global economy: from industrialism to post-industrialism

Until the mid-1960s London was a major centre for light manufacturing. Although the City of London was an important financial centre, and the majority of the labour force worked in services, transport, construction and utilities, 1.4 million people (almost a third of the labour force) worked in manufacturing industry, with more working in the docks and in the port-related processing industries. But, since the 1960s, London has ceased to be a major manufacturing city. In 1967 London's docks were paralysed by the dock workers' strike. Today, it has no dock workers left to speak of and four times as many people work in the financial and business services sector broadly defined, as in manufacturing industry. London has become a post-industrial, informational, city. The notion of 'post-industrialism' initially derives from Daniel Bell's pioneering book *The Coming of Post-Industrial Society* (1973) which argued that modern societies are undergoing a shift from an economy based on the production of physical goods to one based on the provision of services, in which the role of information and knowledge is central. As a consequence of this, Bell argued that education is becoming increasingly important, and that it gives rise to the emergence of a 'new class' of highly educated professionals, managers and technical staff who work in the knowledge industries. Although the idea of a post-industrial economy and society has received considerable criticism since the publication of Bell's book (see Kumar, 1978), the concept of post-industrial or 'informational' cities has become widely accepted (Ley, 1980; Simmie, 1983; Castells, 1989; Hall, 1992; Savitch, 1988). Although there is a debate over the class, social and political structure of post-industrial cities (Walker and Greenberg, 1982), the shift from manufacturing industry towards tertiarisation of the economy, based on services and information processing, has proceeded apace (Noyelle, 1983). Not surprisingly, London's occupational

and earnings structure and housing market reflect these changes. The rapid
decline of manufacturing industry and the growth of financial and business
services has been accompanied by a decrease in the size and importance of
its working classes, and a sharp increase in the size and significance of its
middle classes and their consequent impact on the housing market.

But while London has become a wealthier and more prosperous city during the last thirty years, with a much larger and prosperous middle class, it
has also become a much more unequal city. There is a large and growing
divide between the earnings, the incomes and living conditions of the prosperous expanded middle class and those of the economically inactive, the
less skilled, the low paid and unemployed who have fared far less well.
These changes have been linked to a shift in the structure of the housing
market and in the geographical distribution of different social groups within
the city. As a result, the social geography of Inner London has changed
dramatically in recent decades with major implications for the lives and life
chances of different groups. London has become a more ethnically diverse
city and the gap has widened considerably between rich and poor.

There have also been dramatic changes in the landscape and built environment of London and other global cities, as the structure of their economy
has changed (Ley, 1980; Knox, 1991). As the importance of manufacturing
industry has declined so the physical infrastructure of industrial production
and goods distribution developed in the nineteenth century and the early
twentieth has become increasingly redundant. At the same time, the requirements of post-industrial production and consumption have led to a demand
for new types of space, both commercial and residential.

As Savitch notes:

> Post-industrialism can be seen as a transformation of the built environ-
> ment: factories are dismantled, wharves and warehouses are abandoned,
> and working class neighbourhoods disappear. Sometimes there is re-
> placement of one physical form by another – the growth of office towers
> and luxury high-rises or refurbishing of old waterfronts. Cafés and
> boutiques arise to feed and clothe the new classes.
>
> (Savitch, 1988, p. 5)

Overall, suggests Savitch, post-industrialism

> encompasses a change in what we do to earn a livelihood (processing or
> services rather than manufacturing) as well as how we do it (brains
> rather than hands) and where we do it (offices rather than factories).
>
> (Savitch, 1988, p. 5)

In this book the focus is firmly on London and on the set of linked
transformations which have made it a very different city from that of forty

years ago. But London is not unique and insofar as what is happening in London is similar too or different from what is happening in other major cities, it sheds valuable light on a much wider set of processes and debates. In recent years there has been considerable intellectual debate over the nature and direction of the changes which are believed to be restructuring the major 'global cities' such as New York, Tokyo, Paris, Los Angeles. While most of these cities have undergone major deindustrialisation in recent decades, along with the growth of the financial and business services sector, the implications of these changes for the structure of occupations, earnings and incomes are less clear cut. Although all global cities have become more unequal and socially divided in recent decades, the precise form of the divisions, and the processes generating them, have been disputed. Sassen (1991) argues that New York, London and Tokyo have 'undergone massive and *parallel* changes in their economic base, spatial organisation and social structure' (p. 4), and she suggests that 'transformations in cities ranging from Paris to Frankfurt to Hong Kong and São Paulo have responded to the same dynamic'. While there is no doubt that some world cities have indeed 'responded to the same dynamic', and share a number of characteristics in terms of the structure of their economy, occupations, earnings and housing markets, Sassen's claim is arguably too sweeping in its range. Different cities respond in different ways to the dynamics of globalisation, depending on their existing economic structure, their level of development, their history, position in the global economic and financial system, their social and ethnic composition, and the nature of specific national and local government policies on matters ranging from immigration, labour-market regulation, minimum-wage legislation and wage controls to housing and planning policy. It is not possible to see all cities in any way as clones from a single model. Consequently, their experiences of, and responses to, underlying processes of globalisation and economic restructuring are likely to display a number of similarities and marked differences (Fainstein and Fainstein, 1987; Knox, 1995). This is one reason why we cannot accept the claims of Soja (2000) that Los Angeles represents the leading edge of contemporary urban change. It is one form of many.

Social change in global cities

One major theoretical debate concerning these developments has been associated with the growth of what has been termed 'social polarisation' (Friedmann and Wolff, 1982; Sassen, 1991) and the rise of what have been termed 'dual cities' (Mollenkopf and Castells, 1991) in which the growth of the highly skilled and highly paid financial and business services sector has been paralleled by growth of a low-skilled, low-paid service sector and the decline of the skilled middle-income groups in manufacturing industry. There is little doubt that the size of the professional and managerial groups in global cities has increased in recent years. This is a reflection of the changes which have taken place in the growth of financial, business and other services. The question, however, is whether the growth at the top of the occupational and skill spectrum has been paralleled by major growth at the bottom end, with rising demand for less-skilled service workers to staff the hotels, restaurants, bars, shops, and the host of cleaning and security jobs which have emerged in recent years, or whether the growth at the top end has been paralleled by an absolute contraction across the rest of the occupational structure. In other words, there is a question as to whether London and other global cities have become more middle class or more polarised between, on the one hand, a growing professional and managerial group and, on the other, a large and growing group of less-skilled service workers. There is also an important related question of what is happening to the distribution of individual earnings and household incomes. Are they also becoming more polarised, with more individuals or households at the top and bottom end of the distribution, and fewer in the middle, or are they becoming more unequal, with an increasing gap between rich and poor?

The implications of these two interpretations of what is happening to the occupational class and earnings structure are very different. On the one hand, there are those like Glass who argued that London is becoming more middle class, and that the working classes are thus being squeezed out via the operation of the job and housing markets. On the other, Neil Ascherson (1986) suggests, like Sassen, that while London is undoubtedly witnessing growth at the top, this is in turn causing growth at the bottom end with a growing army of casualised workers. While this could be seen as a narrow academic debate of little relevance or interest to the wider world, this view would be profoundly mistaken. The nature of the changing socio-economic and income structure of London has implications for the structure of economic and social opportunities, the housing market, social segregation, education and for the future of the social order and social conflict. If London is becoming increasingly 'professionalised', and is becoming a city increasingly dominated by the middle classes and high-income groups, it faces a very different future than if it were becoming dominated by increasing numbers of the less skilled and low paid. It is likely that an increasingly 'professionalised' London will be one characterised by substantial increases in demand for owner-occupation in the more attractive areas, rising housing prices, with a spillover into more peripheral areas, which puts more pressure on what remains of the low-income housing market. The analysis put forward in this book of what is happening to the social structure of Inner London is much more closely aligned to that of Glass than to any of the other commentators. It argues that the class structure of London has been characterised by the growth of an increasingly affluent professional and managerial middle class and by the decline of other groups.

Inequality in London

What are the implications of these changes in industrial and occupational structure for the structure of earnings and incomes? The argument made here is that the shift in the industrial structure, from manufacturing to financial and business services, has been linked to change in both labour demand and supply and to a marked upwards shift in the occupational class structure of the city. From a city dominated by a large working class in the early post-war decades, the decline of manufacturing and port-related industries from the 1960s onwards has been accompanied by an equally marked increase in both the size and in the proportion of its white-collar middle class, the great majority of whom work in the service sector. These interlinked shifts in the industrial and occupational class structure have also, it will be argued, been linked to an increase in the size of its high-income labour force and increases in their earnings far faster than those of other groups. Similar changes in the structure of household incomes have led to a marked increase in both earnings and income inequality.

London now has a large and growing middle class, whose earnings and household incomes have grown more rapidly than those of the less skilled, thereby increasing inequality. This growth in inequality is not an accidental by-product of London's role as a global city. On the contrary, it is an inscribed characteristic (Allen *et al.*, 1998): part and parcel of the changing industrial and occupational structure of the city. By skimming off a small percentage of the value of all financial transactions, the financial services industry in almost every western country is able to generate large profits and high incomes for those who work in the sector. Whether we view this in terms of expropriation of surplus value, or efficient markets and the return to human capital, the outcome is much the same: greater wealth and greater inequality.

National differences in occupational and earnings change

The precise form of the transformation of the occupational and income structures of major capitalist countries differs depending on the nature of their national political organisation, the structure and extent of their welfare state regimes and the degree of income redistribution (Esping-Andersen, 1993; Silver, 1993). Although the pressures of economic restructuring, at both the international and the national level, exert a very strong influence on the nature of outcomes, they do not exert a determining influence. There is scope for national and local government policy to influence the structure of employment and distributional outcomes (Hills, 1995, 1996; Atkinson, 1995; Musterd and Ostendorf, 1998; Burgers, 1996). Another factor which influences wage levels between different cities is the potential labour supply.

There are major differences in the structure and price of available labour as a result of immigration policies and the size of the migrant population between Europe and the USA. In New York, Los Angeles and a number of other American migration gateway cities (Clark, 1996), the size of the migrant labour force enables employment of large numbers of low-wage workers in the service sector or in sweated workshops. This is less true of European countries which have more restrictive labour markets and lower levels of migration. As a result, as Esping-Andersen (1993) has pointed out, there is a contrast between countries such as the USA, with large low-wage service sectors, and those European countries such as Germany with more restrictive labour market barriers and generous unemployment benefits but higher levels of unemployment. The degree of polarisation of occupations and earnings seen in the USA will not necessarily be seen in European countries with different welfare state regimes. In these countries, changes in structures of occupation, earnings and incomes are likely to take rather different forms. What the USA and Britain do have in common is a sharp rise in the earnings and incomes of the top decile and, within that, of the top few percentiles. These trends are very strong in New York and London because of the size of the financial services industry.

Although, as this book will show, London and other global cities have witnessed a major increase in earnings and income inequality in recent decades, and a marked shift in their occupational and ethnic composition, the evidence is less clear on the question of social polarisation. It is argued that what is happening in New York and Los Angeles is not necessarily true of London, Paris or Berlin, and that theoretical models imported from North America do not necessarily fit the experiences of western European cities with stronger welfare states and different migration histories. Terms such as social polarisation have a strong resonance and powerful political overtones, but they can mean everything and nothing unless clearly defined. It is dangerous to take the existence of these trends for granted.

The changing structure of London's housing market

The changes in the industrial, occupational and earnings structure of London have been accompanied by an equally dramatic shift in the tenure and price structure of its housing market. In particular, the growth of an expanded middle class of professional and managerial workers, accompanied by substantial growth in earnings and incomes at the top end of the distribution, have increased the pressure of middle-class demand on a housing stock which has grown only slowly. The result, not surprisingly, has been a substantial rise in prices, driven by strong pressure at the top end of the market which has percolated downwards to all sectors of the market.

This has had the result of both making home ownership more difficult for those on average or below-average income, and of increasing the financial incentives for landlords to convert their properties into home ownership. These are characteristics which are shared with many, if not most, other global cities such as New York, Tokyo, Hong Kong and Paris. In all these cities, residential property prices have soared in recent decades and rental property has been sold for ownership. Where they differ, however, is in the extent of their social rented sectors, their degree of rent control and tenant protection and in the extent of planning restrictions on conversions and new building (Murie and Musterd, 1996; Forrest and Murie, 1995). Thus demand pressures from the expanded middle class are mediated in a variety of ways with very different consequences and outcomes. In New York, for instance, there is relatively little social housing and a large privately rented sector. In Amsterdam, by contrast, 70 per cent of housing is social rented which enables a much wider range of groups to live in the city (Musterd and Ostendorf, 1998).

The scale of the housing tenure transformation in London in recent decades has been dramatic. Although home ownership has long been the largest housing tenure in Outer London, Inner London in the 1960s was still strongly dominated by private renting. In 1961 two-thirds of households rented from private landlords and the council sector and home ownership accounted for the remaining third. The last forty years have seen a massive reduction in the size and importance of the private rented sector, and major increases – first, in the size and importance of the council sector, which grew to over 40 per cent of households in 1981, and, since 1980, of home ownership. The 2001 census shows that almost 40 per cent of households in Inner London either own or are buying their own homes. In Inner London an increasing proportion of these homes are conversions from privately rented houses and other buildings. Much of the rest of the privately rented sector has disappeared as a result of large-scale local authority clearance and redevelopment of high-density and high-rise council estates in the 1960s and 1970s which transformed the residential environment of much of Inner London. Whilst this provided a massive improvement in internal housing conditions, the overall design of many estates has subsequently proved both physically and socially very problematic.

Accompanying this transformation in the housing tenure structure of London, there has been a social reshaping of the major tenures. Until the 1970s, council housing in London was dominated by the skilled working class and junior white-collar workers. Since then, an increasing proportion of what remains of the skilled working class have either moved out of London or moved out of council housing into home ownership. In the process, what remains of the council sector has become increasingly dominated by the less skilled, low paid, unemployed, the economically inactive, single parents and some ethnic groups. It has effectively become a socially residual tenure. Conversely, home ownership has now become the dominant tenure for the majority of white-collar and skilled-manual households, and a process of socio-tenurial polarisation has taken place which has seen the population of London increasingly segregated by economic activity, employment, occupation, income and ethnicity into the two distinct tenures.

Changes in ethnic composition

The ethnic dimensions of employment and housing market change are important as the ethnic minority population of London has grown dramatically in recent decades. The 2001 census showed that almost 30 per cent of London's population, and over a third of the population of Inner London, consisted of ethnic minorities, and the proportion looks set to rise both as a consequence of the age structure and high birth rates of migrants. The composition of recorded migrant flows to London has changed in recent years, however. Although an increasing proportion of migrants are refugees and asylum seekers, many of whom have little or no English, the number of skilled international migrants from developed countries has risen, reflecting the increasing flows between such countries. While some ethnic groups, notably the Indians and the Chinese, have rapidly acquired educational qualifications and have moved into skilled white-collar jobs, with many professionals, other groups, notably those of West Indian, and Bangladeshi ethnic origin, are still trapped in a cycle of discrimination, poor educational attainment, low skills, unemployment, low incomes and council housing.

Gentrification and segregation

One consequence of the growth of the professional, managerial and technical middle class is that they have outgrown their traditional residential areas in Inner London and have expanded outwards into what were previously working-class areas. This process of 'gentrification' is very well known, but it has grown enormously in importance in the 1980s and 1990s and the growing middle classes are now pushing out into areas of Inner London such as Hackney, Tower Hamlets and Vauxhall, which were, twenty years ago, still bastions of the working class. This process has been accompanied and partly caused by the almost relentless rise in London house prices from an average of around £5,000 in 1970 to £220,000 in 2002.

In market economies, where housing is allocated through the price mechanism on the basis of ability to pay, an increase in the size and/or the purchasing power of highly paid groups means they will almost invariably outbid other groups in the competition for scarce desirable resources, not least for attractive housing in attractive areas. Ruth Glass suggested an almost Darwinian class struggle for housing, writing that:

any district in or near London, however dingy, or unfashionable, is likely to become expensive, and London may quite soon be a city which illustrates the principle of the survival of the fittest: the financially fittest, who can still afford to work and live there.

(Glass, [1963] 1989, pp. 140–1)

Forty years on, Glass's predictions have proved uncannily correct. As house prices have soared, the middle classes have expanded into hitherto working-class, inner-city areas in their search for accessible and affordable housing. Inner London has been gentrified with a vengeance.

A result of these changes in the social and spatial structure of the housing market is that Inner London has simultaneously become less socially segregated at the macro scale and arguably more segregated at the micro scale. There is no longer a simple divide between a relatively small middle-class West End, and a large working-class East End. Instead, as the middle classes have pushed out in hitherto working-class areas, the gross segregation between the working-class Inner London boroughs and the middle-class areas has been reduced. But, simultaneously, there is now arguably an increased micro-scale segregation, sometimes on opposite sides of the street, between relatively well-off middle-class owners living in renovated or converted period houses and low-skilled, low-income council tenants. The implications of these changes for the lives and the opportunities of London's population in terms of employment, incomes, housing and education are considerable.

Economics and politics in the reshaping of London

The focus on changes in the structure of London's economy in driving changes in other aspects of the city raises important questions regarding the relative importance of economics, politics and planning in shaping urban development. This question has been explicitly discussed by Fainstein (1990) in her comparison of London and New York in the 1980s. She argues that while there is no doubt that economic forces were the primary forces shaping the two cities, there were alternative paths which policy-makers can follow involving more or less citizen participation and redistribution of the growth dividend. She argues that the governments of the two cities opted for 'the route of minimal planning, dampened participation and market distribution', and that these politics have contributed to highly uneven development, shortage of affordable housing and the like. Although she concedes that given the capitalist framework and economic situation of the two cities, government could not have created a very much different form of development (p. 553), they could have done more to mitigate some of its negative effects.

Fainstein is correct. It is difficult to escape the conclusion that the fundamental forces driving the transformation of London in recent decades have been the changes in the structure of its economy and its occupational and earning structure rather than (with the possible exception of the regeneration of Docklands) politics and planning. This is not to suggest that politics and planning have been unimportant, but that the primary driver of change has been economic. The demise of dock work in London is a result of changes in transport technology, including a move to larger ships, containerisation, and decisions made by the Port of London authority and others to shift port activities downstream to purpose-built facilities. So too, the closure or decentralisation of manufacturing from London, and the associated decline of manual jobs, reflects corporate decisions to close down and/or seek lower-cost production sites elsewhere in Britain or abroad.

Although there have been attempts by local government and planners, notably under the Labour-controlled GLC, to resist deindustrialisation, culminating in the London Industrial Strategy (GLC, 1985b) and the formation of the Greater London Enterprise Board (GLEB), in most cases their impact was relatively marginal. A few thousand manufacturing jobs may have been saved or created, but hundreds of thousands of jobs were lost in the 1970s and 1980s as they were in other cities and other western countries. Whilst it can be argued that the GLC was struggling to implement socialist policies in the face of strong political opposition from a Conservative-controlled central government determined to limit the power of Labour-controlled local governments, it is difficult to see how the GLC's industrial policies would have been more successful under a Labour national government.

Abolition of the GLC in 1987 by Margaret Thatcher ensured that there was no overall strategic local government body for London (O'Leary, 1987; Toulouse, 1992) until the creation of the Greater London Authority in early 2000. For most of the period the London boroughs have simply reacted to changes in the economic environment, with limited ability to influence the direction or pace of change. The most they have been able to do has been to use their planning powers to try to slow or halt proposed land-use changes which they believe to be detrimental to London and Londoners.

This has been very effective in some cases, for example in slowing the conversion of the West End into offices, but although local government can put forward initiatives to try to help the regeneration of decaying or derelict areas, it has very little power to ensure that such development takes place. Instead, it is largely dependent on attracting private sector investment, and this is only likely if the development will be profitable. The private sector does not knowingly undertake unprofitable investments, however socially laudable or desirable they may be, and the main options for local government are often to try to oppose private development or to work with it. Since the creation of the Greater London Authority, it seems the Mayor of

London Ken Livingstone has committed himself to strengthening London's role as a major international financial centre, rather than trying to resist it. In this respect, the GLA is working with the grain of recent changes in the structure of London's economy from manufacturing industry to financial and business services, rather than against it. Ken Livingstone has made several statements stressing the need to ensure the development of London's built environment to accommodate the space needs of financial institutions.

The role of neo-liberal policy

Against this interpretation it can be argued that the dominance of market forces in the reshaping of London in recent decades is neither inexorable nor inevitable but is, rather, the result of the policies followed by the Conservatives since 1979 under Margaret Thatcher's leadership which privileged market forces and deliberately weakened the role of the local state, the unions and the working classes. Toulouse outlines the global restructuring as follows:

> In attempting to model the dynamics of the new era, theorists have focused on the increasing mobility of capital and the leverage transnationals and international capital markets can now exert on state policy-making at all levels. In post-Fordist terms, capital is migrating toward a new regime of accumulation, and nation-states and localities are scrambling to foster modes of regulation appropriate to particular niches in the global circuit of capital. When classes appear in these accounts they are usually in the expiring form of the old industrial working class . . . or in the emerging form of an international upper class and a professional middle class . . . In short . . . we may say that there is a general tendency to conceive classes as casualties or as emerging consequences of the restructuring process imposed by the globalization of capital.
>
> (Toulouse, 1992, p. 56)

Toulouse is highly critical of this interpretation and argues, instead, that classes and class politics are a cause as well as a consequence of social change in global cities. More specifically, he argues that 'much of what happened in London was caused by the Thatcher government's attempt to use state power to shape the social polarization wrought by economic restructuring to class based ends' (p. 56). In making this claim, Toulouse explicitly seeks to avoid the assumption that 'the Conservative party had a conscious class-based intent to foster the long-term interests of capital' (p. 57). Instead, he argues that Thatcherites had no master plan for the economy based on the logic of capital, but rather 'a set of general priorities based on the aspirations of the suburban middle class and a steadfast determination to push small-scale changes toward long-term political gains' (p. 57). Toulouse argues that the focus of attention should not be the intentions of Conservative urban policies, but their strategic effects, and he points to the spatial effects of the Conservatives' economic strategy to modernise the British economy. By working with the trends of economic restructuring rather than trying to shore up manufacturing areas, 'the strategic effect was to abandon all manufacturing that could not compete internationally and to compel cities to specialize in corporate services and high income consumer markets' (p. 58).

There is considerable validity in this interpretation, and Hudson and Williams (1995), Johnston (1979), Jessop *et al.* (1988) and Peck and Tickell (1992), all argue that Thatcherite policies, including deregulation, privatisation, interest rates, and higher-rate income tax cuts, benefited the private sector, financial services, the middle classes, London and the South East at the expense of the public sector, manufacturing, the old industrial regions and the working classes. Analysis of the impact of the Conservatives' higher-rate income tax cuts (Hamnett, 1997) shows the major benefits which accrued to high-income groups, who are strongly concentrated in London and the South East. There is no doubt that there was a clear political policy to weaken local government, the public sector and trade unions and to strengthen the role of the private sector and the middle classes, and that the effect of high interest rates weakened the competitiveness of manufacturing industry. It is more difficult to argue that there was an explicit policy to benefit London and the South East. But, as Peck and Tickell (1992) point out, the inflated economy of the South East was driving national economic policy, both in the boom of the late 1980s and the subsequent slump. 'In this way a crisis of regulation within the core region effectively triggered a national accumulation crisis' (p. 359).

It can also be argued that the ability of local governments to act is also constrained by central government policy, and the Conservatives' creation of the London Docklands Development Corporation placed the most important development in London entirely outside the framework of local government control. It is important to note, however, as Stevenson (1998) stresses, that the LDDC was very much 'a creature of the state'. It may have acted largely autonomously of local government, but its creation, brief and *modus operandi* were shaped by state policy. As Stevenson comments:

> One of the great ironies of recent times in the history of urban politics is that a new right wing government, committed vociferously to giving free reign [*sic*] to market forces, created an enormously powerful state institution (the UDCs) to intervene in the land market adjacent to the largest city centres in the country and poured billions and billions of public money into them. The double irony was, certainly in London,

that it was the collapse of the private market in land amid the recession of the early 1990s that contributed decisively to altering the role and mode of operation of the biggest UDC of them all – the LDDC.

(Stevenson, 1998, p. 24)

But, considering Thatcherite policy as a whole, it is clear that Conservatives saw one of the key roles of government as being to underpin the growth of the free market and rolling back the state (Thompson, 1990; Feigenbaum *et al.*, 1998). If this meant the paradox of greater state intervention in order to strengthen the market and weaken the state (as with privatisation), so be it. In addition, Conservative policy to deregulate foreign exchange controls in 1979 and to deregulate the City of London in 1986 clearly set the framework for the expansion of the City in the mid–late 1980s to take place. National government policy sets the parameters within which markets can operate, and this both underpinned and encouraged the transformation of London over the last twenty years. To stress the role of economics in the transformation of London is not, therefore, to argue that the state has played no role, but to argue rather that the state (in the shape of central government) set the key parameters which have allowed market forces to operate more freely. The Conservatives, and the subsequent Labour government, could have retained the interventionist regional and industrial policies which characterised the Wilson government, but they explicitly chose to introduce a set of policies which gave market forces a much greater role.

The impact of government policy can also be seen in terms of the growth and decline of the council sector over the last forty years. During the 1960s and 1970s a massive council house building programme took place, both in Inner London and elsewhere, encouraged and largely financed by central government. But, with the election of the Conservatives in 1979, the introduction of the tenant Right to Buy legislation and the sharp cutbacks in the funding for new council building, the period of major council building rapidly ground to a halt. Subsequently, the dominant force in the London housing market has been the growing demand for housing from the professional and managerial classes, the rise in prices this generated, and the consequent opportunities for private landlords to sell out to owners and for developers to initiate conversions of one kind or another. Local authority planners do have the ability to turn down planning applications that do not fit the structure of local plans, but they cannot ensure socially desirable developments. Most key decisions on the evolving shape of the economy and its land-use implications are taken elsewhere, and planners and local authorities have limited scope to respond. Firms cannot be compelled to stay in business, or conjured up out of thin air, though it may be possible to attract them to an area if the market and infrastructure are right. In many respects, central government policies on financial deregulation in the early 1980s, abolition of wages councils, cuts in income taxes and the establishment of the London Dockland Development Corporation, have been more important than local ones in influencing the directions of change, though in some Conservative-controlled flagship boroughs such as Kensington and Wandsworth councils have been able to intensify the pace of gentrification through their housing policies. They were able to do this, however, because their policies were congruent with those of central government, rather than opposed to them. It is difficult to argue, as Toulouse does, that central government policies have been primarily responsible for the direction of the changes which have taken place in London. But he is quite correct to argue that 'Class based parties utilize the market making powers of the state to shape economic restructuring to their own ends' (p. 56), and Conservative policies have clearly worked in favour of capital and the middle classes.

Structure of the book

The structure of the book is straightforward and mirrors the argument presented here. Chapter 2 briefly outlines theories of world and global city development, examines the role of London in the international economy, and looks at the changing structure of London's economy, particularly the decline of manufacturing employment and the growth of financial and business services. It argues that the economy of London has changed from that of an industrial city in the 1960s to that of a post-industrial service city today. While this has some advantages, it also brings costs in terms of the type of jobs which are available and demand for different skills. Manufacturing workers have suffered particularly badly in this respect. Chapter 3 examines the various theories regarding changes in occupational class structure in capitalist societies, particularly in global cities, and sets out the three main competing interpretations. It also provides an empirical study of changes in London's socio-economic structure from 1961 to 1991, arguing that the key trend is professionalisation not polarisation or proletarianisation. Chapter 4 looks at the changes in the structure of earnings and incomes in London and compares them to Britain as a whole. It focuses on the question of whether these structures have become more polarised or more unequal, and shows that the answer is clearly the latter. Not only are there more high earners in London than ever before but their earnings have grown much faster than those of other groups. It also examines the structure of earnings in the City and various business services, looking at evidence for the growth of high earnings. Finally, it looks at the changes in household incomes in London and shows that these too have become far more unequal, particularly as dual-earner professional and managerial households have become more important and the number of unemployed, benefit-dependent households has risen.

Chapter 5 looks at the role of migration in London, focusing both on the changing structure of migrants over time, the growing importance of professional migrants who come to work in financial and business services, the growth of ethnic minorities and the change from a mono-ethnic to a multi-ethnic London. It also looks at the problems facing some ethnic minorities in terms of both their labour market and housing market position. It shows that while some groups have done well and moved into professional and managerial jobs, others have concentrated in both low-skilled areas of the labour market and into council housing. Finally, it looks at the geographical distribution of ethnic minorities in London, and asks if they have become more or less concentrated. Chapter 6 turns to look at the impact of these developments on the housing market in London, showing the dramatic transformation of the housing tenure structure from private renting in Inner London – first to local authority housing, and, more recently, to ownership. It examines the various forms of transformation, including conversions of houses into flats, and commercial property to residential. It shows the emergence of the growing social divide between the two main tenures in London, and it focuses on the changes in house prices and housing affordability which have meant that only the more affluent can afford to buy. Chapter 7 shows the implications of the changing occupational and house price structure of London for development of gentrification. It argues that the growth of an expanded professional and managerial middle class, many of whom work in central or Inner London, has put increasing pressure on the private housing market and has led to the rapid residential expansion of the middle class into hitherto working-class areas of London. As prices have risen, so many working-class residents have either sold up and moved out, or been directly or indirectly displaced. Chapter 8 examines the implications of some of these changes for issues of social inequality, social segregation and social exclusion in London. It argues that London exhibits the paradox of great wealth and considerable poverty in close juxtaposition. Less-well-off residents are concentrated in areas with poor schools, low educational attainment and poor health, but the impact of gentrification has if anything intensified the divisions between well-off homeowners and the economically inactive, unemployed or low-paid council tenants, at the local level.

Finally, Chapter 9 examines the changing landscape of London. It argues that as the importance of manufacturing industry has declined and that of financial and business services has increased, that the physical landscape of London is being transformed from an industrial to a 'post-industrial' city. This is seen both in the disappearance of the docks, the expansion of the City of London, the emergence of Canary Wharf, and in the transformation of various types of industrial land use such as canal basins, railway goods yards and the like, into new office and residential areas. Many of the old warehouses and factories in Inner London have been or are being converted into luxury apartments for the expanded middle class, particularly along the river Thames and in the old East End, which is now becoming a fashionable area to live. At the same time, the redevelopment of large areas of poor-quality rented housing in the form of high-density and often poorly designed council estates has dramatically changed the landscape of parts of Inner London, often much for the worst for its residents.

Chapter 2

The economy of London

From industrial to post-industrial city

London is the capital of capitals; the head which directs the hand of labour on a national and international scale. As a trading centre, subsequently as a centre of finance and a market for services as well as a centre of government, London has always been a centre of information and control.

(Greater London Council, 1985b, p. 334)

The early 1980s marked the beginning of a new stage in the development of London and New York. While other economic sectors were dispersing geographically, certain advanced service industries centring around financial activities intensified their presence in the centres of these cities. This intensification resulted primarily from the enlarged role of financial capital in co-ordinating the world economy and the extremely active deal making that accompanied this role. The decade witnessed the birth of new financial markets for the exchange of arcane financial instruments and the raising of huge pools of credit to underwrite speculative activities in property development, mergers, and leveraged buy-outs, as firms themselves became negotiable assets.

(Fainstein and Harloe, 2000, p. 155)

Introduction

The economy of London has undergone a dramatic transformation over the course of the last thirty to forty years. Until the mid-1960s it was a major centre of British light industrial production, and a third of its labour force were employed in manufacturing. Although it had long functioned as an important national and international financial centre, the importance of finance and business services for overall employment was relatively small until the 1980s. In 1961 only about one in ten of all London's workers were employed in this sector. But in the last forty years the proportionate importance of manufacturing industry and finance and business services has been totally reversed. In 1998, finance and business services employed almost a third of London's workers, whereas manufacturing industry employed about

8 per cent. This chapter looks at the nature of this transformation from an industrial to a post-industrial economy and the structure of London's economy at the end of the twentieth century.

One of the major arguments in this book is that in order to understand the changes which have transformed London over the last thirty years it is necessary to understand London's changing position in the world economy and financial markets. As capital of the British Empire London has been a major centre of international finance since the early nineteenth century (King, 1990). But in the last thirty years its role as one of the big three international financial centres has reinforced its international importance. It is generally accepted (Coakley, 1992; Thrift, 1994) that the current prosperity of the City of London as one of the three key international centres of finance, along with New York and Tokyo, rests on two principal factors. The first was development of the Eurodollar market in the 1960s, and the second was the deregulation of the City in 1987: the so-called 'Big Bang'. What this did, amongst other things, was to break down the traditional institutional barriers between jobbers and stockbrokers and the barriers to international ownership of London-based firms (Auger, 2000). By getting rid of these restrictions, the City opened itself up internationally to the development of integrated financial insitutions. With the added advantage of the English language it rapidly became the premier international centre for foreign exchange dealing and international stock trading and fund management (Davis and Latter, 1989; Lee and Schmit-Marwede, 1993; Coakley and Harris, 1983). As Toulouse (1992) comments: 'The City is now a global stage located in Britain, rather than a British stage in the global arena' (p. 62).

Major cities have long played a key role at the centre of the world's economic, trading and financial systems: ancient Rome, Renaissance Florence and Venice, seventeenth- and eighteenth-century Amsterdam. But in 1795 the French occupation of Amsterdam ended its period of international domination and over the next twenty years almost continual warfare on the Continent, ending with the Battle of Waterloo in 1815, enabled London to emerge as Amsterdam's successor. London's dominance went largely unchallenged through the nineteenth century and the early twentieth. Until the First World War London, Paris and Berlin were the centres of their respective empires, controlling half the globe between them (Kynaston, 1994). As Dyos put it:

> London had already become, not merely the country's capital, but the centre for most of the big commodity markets and financial houses, the country's chief port and industrial area. This commercial metropolis was being made more impregnable still by the concentrative power of the railways at home and the gains of economic imperialism abroad.
>
> (Dyos, 1967, p. 28)

And Charles Booth noted of London in typically florid imperialistic prose:

> It is not only an unrivalled national emporium and world market, but it is also the mother city of the Kingdom and of the Empire. London is the centre, moreover, not only of the Imperial government and the Judiciary, but also of banking and finance, both national and international. It is in London that the agent's general in the great colonies, as well as the chief, business agencies, and official commercial representatives of foreign countries are found, their presence illuminating the fact that it is the recognised national centre, not only of government, but of trade. Everything can be bought in London and therefore everyone comes to buy ... London is as much an emporium for raw materials coming from all over the world as for finished products.
>
> (Booth, [1901] 1982)

Booth's description is typical late-Victorian bravura, but the points he makes about the role of London as the centre of the Empire and trade are valid and, despite the end of the Empire, the key role of London and a number of other major cities such as New York, Frankfurt, and Toyko in business and finance has intensified rather than weakened in recent years. This raises a series of questions about why these cities have become so important and their role in the world economy. This issue is discussed below before turning to look in detail at the fortunes of different sectors of London's economy.

The role of world cities in the global economy

Various writers have grappled with the definition of world or global cities in recent decades. One of the first academics to attempt to define the characteristics of these cities was Peter Hall in his book *The World Cities*. Hall suggested that:

> they are usually the major centres of political power. They are the seats of the most powerful national governments and sometimes of international authorities; of government agencies of all kinds. Round these gather a host of institutions, whose main business is with government: the big professional organizations, the trades unions, the employers' federations, the headquarters of major industrial concerns. These cities are the national centres not merely of government but also of trade. Characteristically they are great ports, which distribute goods to all parts of their countries, and in return receive goods for export to the other nations of the world ... Traditionally, the world cities are the leading banking and finance centres of the countries in which they stand. Here are housed the central banks, the headquarters of the trading

banks, the offices of the big insurance organizations and a whole series of specialized financial and insurance agencies.

(Hall, 1966, p. 7)

Hall goes on to note that while 'government and trade were invariably the original *raison d'être* of the world cities' (pp. 7–8) they quickly became the centres of professions, the law, museums, press and television; 'places where information is gathered and disseminated'. His characterisation of world cities as centres of political, economic and financial *power* is important, and serves to distinguish the world cities from other cities which, however large in terms of size or population, or important in national terms, do not occupy a powerful position in the world economy. But in retrospect it seems overly descriptive and atheoretical. In addition, the linkages between political, economic and financial power were never as close as he suggested. While this linkage is historically true of London and Paris and pre-war Berlin, it was never true of the USA, where New York and Washington occupy quite distinctive roles. Frankfurt has also become the most important European financial centre outside London, despite its lack of formal political power.

Friedmann and Wolff's world-city hypothesis

More recently, however, a more explicitly economic and financial interpretation has been put forward to account for the power of the world's major cities. It argues that as a result of the growing internationalisation of the world economy a new type of city – the world city – has come into being which performs a different role to the imperial cities of old. These cities are the control and command centres of global capitalism, and Friedmann and Wolff (1982) and Friedmann (1986) argued that the world economy is now controlled by transnational corporations that have increasingly freed themselves from national constraints and have organised global production and markets for their own purpose. They see different localities as being enmeshed within the global system in a variety of ways:

> The specific mode of their integration with this system gives rise to an urban hierarchy of influence and control. *At the apex of this hierarchy are found a small number of massive urban regions that we shall call world cities.* Tightly interconnected with each other through decision-making and finance, they constitute a worldwide system of control over production and market expansion.
>
> (Friedmann and Wolff, 1982, p. 310; emphasis added)

They argued that although the concept of a world economy articulated through key cities is not new, the present situation is different because the world economy is far more integrated and interdependent than in previous eras. In their view, the emerging new transnational global economy is in principle unlimited and is controlled from what Wallerstein termed the core areas of the global economy: the older, already industrialised and possibly post-industrial regions that contain the vast majority of corporate head-quarters and are major markets for world production. As they put it:

> Even imperial London, ruling over an empire 'where the sun never sets', controlled only portions of the world. The present transnational system of the space economy, on the other hand, is in principle unlimited . . . The world economy is no longer defined by the imperial reach of a Rome, a Venice or a London, but by a linked set of markets and production units organised and controlled by transnational capital. World cities are a material manifestation of this control, and they occur exclusively in core and semi-peripheral regions where they serve as banking and financial centres, administrative headquarters, centres of ideological control . . . Without them, the world-spanning system of economic relations would be unthinkable.
>
> (Friedmann and Wolff, 1982, pp. 311–12)

For Friedmann and Wolff, world cities play a dual role, lying at the junction between the world economy and the territorial nation-state. They point out that there is nothing inevitable about either the structure of the world economy or its exact materialisation in world cities, and that the driving forces of competition, the need for accumulation and polit-ical struggles make the intersection of the world economy and world cities a point of intense conflict. Nonetheless, they put forward the central hypo-thesis that:

> the mode of world system integration will affect in determinate ways the economic, social, spatial and political structure of world cities and the urbanising processes to which they are subject.
>
> (Friedmann and Wolff, 1982, p. 313)

Friedmann and Wolff established a new field of inquiry: the relation-ship between the development of transnational companies, the world eco-nomy and the urban system required to control such an economy. Their work, which was strongly influenced by Wallerstein's idea of a capitalist world economy, has proved extremely influential. However, not only did it fail to explain *why* world cities should be home to the headquarters of large transnational corporations, some of which are in fact moving their headquarters out of such cities (Lyons, 1994; Lyons and Salmon, 1995; Marshall, 1993), it also paid little attention to the role of finance and business services.

Cities as centres of financial and business services

This issue was taken up by Cohen (1981), who argued that the importance of world cities lay in their role as centres of corporate services. He argued that, with the rise of a new international division of labour post-war, and the development of multinational corporations, the operations of such firms are now spread out across the globe. This requires information about potential markets, and greater control over investment and production, distribution and sales worldwide. While routine decisions are delegated to local managers, it is necessary to maintain a greater degree of central control and co-ordination over worldwide operations to ensure continuing success in developing new products and markets. Cohen placed considerable stress on the changing structure of the major companies, in particular their widening geographical spread and the growing complexity of the international business environment – both of which require a greater need for, and reliance upon, sophisticated or specialist commercial services such as international law and accountancy firms, management consultancies and the like. Such firms have grown rapidly in size and importance in the last thirty years, and they are overwhelmingly concentrated in the major world cities (Daniels, 1991; Thrift, 1987; Lyons and Salmon, 1995; Moss, 1987; Warf, 1989; Taylor and Walker, 2001).

Cohen argued that while these services have provided support for business operations in the past they have now become key sources of information and intelligence which are of crucial significance to the global operation of companies. He argues that similar changes have occurred in corporate dealings with governments and other regulatory bodies, that big law firms now play a major role in dealing with political and legal issues across the globe, and that this has significant implications for the corporation as a whole. As a consequence of these demands, there has tended to be a concentration among these commercial services into a smaller number of large firms which are now overwhelmingly based in a small number of major cities. Thus business services have become more concentrated at the top of the urban hierarchy (Lyons and Salmon, 1995).

There is no doubt that worldwide operations do require more sophisticated financial, legal and accountancy inputs than purely national operations, and the concentration of advanced business services in the major cities suggests that this is where the market for such services is greatest. Indeed, the international distribution of US advertising agencies, the European and Asian offices of major US law firms and international accountancy firms, are almost all concentrated in the major world cities (Beaverstock et al., 1999, 2000; Taylor, 2000). The distribution of these offices provides a map of world city status in certain key respects.

Sassen's global city thesis

More recently, in her influential book *The Global City: New York, London and Tokyo*, Sakia Sassen (1991, 2001) developed and elaborated these ideas, arguing that the global reach of the world economy necessitates more centralised control and co-ordination, not less. As she puts it:

> a combination of spatial dispersal and global integration has created a new strategic role for major cities. Beyond their long history as centres for international trade and banking, these cities now function as centres in *four* new ways: *first* as highly concentrated command points in the organization of the world economy; *second*, as key locations for finance and specialized service firms, which have replaced manufacturing as the leading economic sectors; *third*, as sites of production, including the production of innovations, in these leading industries, and *fourth*, as markets for the products and innovations produced.
>
> (Sassen, 1991, p. 3; emphasis added)

She goes on to argue that as a result of these changes in the structure of the global economy and the functioning of major cities, a new type of city has arisen: the global, of which London, New York and Tokyo are key representatives. While Sassen agrees with Friedmann that global cities function as command points for the global economy, she differs in the emphasis she gives to financial and commercial service production and innovations. Her focus is not the power of the large multinationals to co-ordinate and control the global economy, but rather upon the *practice* of global control. By this, she means the production of those services – financial and commercial – which actually do the controlling and managing across the corporate networks. It is in this sense that global cities produce things: they are sites of production for services and financial products which co-ordinate global economic activity. Sassen argues that London, New York and Tokyo have evolved into strategic sites for the development, production and supply of innovative financial and business services. These services have permitted, and have simultaneously benefited from, the dispersal and fragmentation of manufacturing and, more recently, certain types of routine services, combined with a globalisation of capital transfers and corporate management. This development, she suggests, started in the 1970s but accelerated substantially with widespread financial deregulation during the mid-1980s.

More recently Thrift (1994) argued that cities are primarily *centres of interpretation*. What he means by this is that their importance lies not simply in their role as centres of financial products and business services, but as places where information about the global economy, companies and financial flows is exchanged and interpreted. Walter Wriston, the ex-chairman of Citibank, has famously commented: 'Information about money is now almost as important as money itself.' Thrift argues that the City, like other key financial centres, functions as a centre of knowledge and expertise in four related ways. First, it functions as a nexus of face-to-face contacts for the exchange of information. Second, it's a centre for electronic information gathering and transmission. Third, it is a centre of textual interpretation, including the financial press and media. Finally, argues Thrift, the City is

> increasingly home to many different global 'epistemic communities', occupational communities with their own specialized vocabularies, rhetorics, knowledges, practices and text. From economists to foreign exchange dealers to Eurobond traders each of the communities tends to live in an increasingly specialized narrative world.
>
> (Thrift, 1994, p. 350)

Thrift argues that in this respect the City increasingly markets itself as a 'centre of cultural authority' for global financial services, a place where people from around the world come to exchange knowledge and information, trade and deal. His answer to the question why the City has not simply dissolved in the electronic ether is that, paradoxically, the need for face-to-face contacts has grown rather than diminished. As he puts it: 'electronic communication seems to have acted as a supplement to face-to-face communication, rather than an alternative, increasing the overall level of communication . . . into and out of the City' (1994, p. 351).

It is important to add one further point. Much of the literature in the 1980s and early 1990s regarding the role of global cities focused almost exclusively on their economic and financial role. But, as Friedmann and Wolff (1982) pointed out, these cities also have another key role as centres for the production, transmission and consumption of culture. This function was largely obscured until the late 1990s when two authors in particular, Sharon Zukin (1995) and Alan Scott (1997, 2000, 2001), argued that culture is increasingly a key component of the business of cities. Scott, in a series of articles, has suggested that in late twentieth-century capitalism, the symbolic role of goods and services has become evermore important, and that global cities function as the centres for the production and circulation of cultural images and meanings on a global scale. They are centres of the creative and cultural industries (Hall, 2000), and the role of Los Angeles as a global centre for film production and distribution is illustrative of this – but it is characteristic of all global cities and Pratt (1997), for example, has shown that London contains a disproportionate concentration of what are termed the 'cultural industries' such as film, video, radio, TV, music and advertising and a wide variety of other activities such as website design (Greater London Authority, 2002).

London: from industrial to service economy

London has been a major international financial and business centre for almost three hundred years; from the early eighteenth century onwards it began to take over from Amsterdam, through the rise of the British Empire in the nineteenth century and the early twentieth and the associated growth of the trade, shipping, finance and insurance businesses. Lloyd's, the London Metal Exchange, and Baltic Exchange all established their position. London, by virtue of its role as capital city, has always had a large service sector and it has long been a centre for luxury goods production and consumption associated with the court, the London Season and its general role as a centre of wealth, power and splendour. In the eighteenth and nineteenth centuries it was the main centre for clock and watch production, printing, dressmaking, tailoring and other luxury goods bought by the landed aristocracy and the gentry (Hall, 1966). The nineteenth century also saw development of its banking and financial role as a centre of Empire (Jacobs, 1994; Kynaston, 1994; Mitchie, 1992; Thrift, 1994). As Stedman Jones notes:

> Historically, the economic importance of London depended upon three closely related factors: firstly, it was the major port of the English import and trans-shipment trade; secondly, it was by far the largest single consumer market in England; and thirdly as a centre of government and the court, it was the focal point of conspicuous consumption and its attendant luxury trades. In the period before the Industrial Revolution, because of the predominance of handicraft production, and primitive transport facilities, these factors encouraged the growth, not merely of finishing and consumer trades, but also of semi-processing and capital goods industries like leather and sugar manufacture, shipbuilding and silk production. The closeness to market, the access to raw materials, the close interrelation of city and government, and the presence of a highly skilled labour force gave London an impressive industrial advantage.
>
> (Stedman Jones, 1971, p. 20)

The nineteenth century and the early twentieth saw the development of London as one of the major, if not the major, industrial centre in Britain. Although it was never a centre for heavy industry, its role as a major port and the associated port-related processing industries, as well as the development of light industry and consumer goods in the 1920s and 1930s, established it as a manufacturing centre. The major role of London in light engineering and other trades up until the mid-1960s has been well documented by Hall (1964) and Coppock (1964). The inner areas of London, adjacent to the City, contained an extensive variety of light industry such as furniture making, instrument making and printing and publishing, and London's role as a centre of light industrial production grew considerably during the inter-war years. As White points out:

> London was ... the beacon illuminating the nation's path out of depression, the forcing ground of new ideas, new industries, new ways of living. The image had its material grounding in the glass and concrete industrial suburbs of the Lea Valley and West Middlesex. London and its fattening satellite towns were home to the fastest growing industries of Britain, a growth at its most dramatic between 1925 and 1935: radios, batteries, vacuum cleaners, electric lamps, radiant fires, extractor fans, meters and photographic equipment; motor cars, buses and vehicle parts, ladies' nylons, potato crisps, celluloid dopes, artificial limbs – all made in London, now more than ever, the capital of finished goods manufacture.
>
> (White, 1986, p. 36)

This change resulted in substantial employment growth in London in the 1930s when other parts of Britain were still experiencing high levels of post-slump unemployment.

But, in recent decades, London's economy has undergone a dramatic transformation from an industrial city to a post-industrial city dominated by financial and business services in which manufacturing industry and employment has been reduced to a vestigial remnant of its former importance. From being one of the foremost light manufacturing centres in Britain, London has been comprehensively deindustrialised and its industrial structure has changed from manufacturing to the financial, business services and creative industries. London no longer makes 'things', in the traditional sense of manufactured goods; rather, it now provides specialist services and produces financial and creative products ranging from legal advice, computer software, music, TV and video programmes, websites, financial derivatives and other intangibles. In many respects it is now primarily a 'weightless economy' (Coyle, 1997; Quah, 1997).

In 1971 London had 86 of the biggest 100 British company headquarters (Westaway, 1974; Evans, 1973). Although it still possesses a disproportionate share of headquarter offices of Britain's major companies, this role is primarily one of control and co-ordination, and most physical production takes place outside London.

Because of its leading role as major international financial centre, London contains a disproportionate concentration of jobs in financial services. Secondly, because of its role as the international headquarters of major corporations it contains a high proportion of jobs in corporate headquarters and in related business services such as management consultancy, law, advertising, design and public relations. Third, because of its role as a centre of broadcasting and publishing, it has a disproportionate concentration of jobs in media production and marketing, ranging from national newspapers and journals to the major TV studios, record and video production. Finally, particularly in the 1990s, it has become a major centre for the cultural and creative industries and the new digital economy. Like Paris and New York, London is now a key centre for ecommerce, website design and the like. Before looking at the growth of these sectors it is important to examine the decline of manufacturing industry in London and the transformation of its employment structure in recent decades.

The decline of manufacturing industry

London's decline as a manufacturing centre has been comprehensively documented (Buck *et al.*, 1986). The period from the mid-1960s onwards has witnessed a dramatic contraction of London as a manufacturing centre. In 1961 London had 1.45 million manufacturing jobs (32.4 per cent of the total) in electrical engineering, food, drink and tobacco, chemicals, instrument engineering, paper and printing, furniture making, clothing and footwear. By 1981 it had fallen by just over 50 per cent to 681,000 (19 per cent of the total).

The changes in employment structure in Greater London from 1981 to 1991 (Table 2.1) reinforced the changes seen in previous decades. First, total

Table 2.1 The employment structure of Greater London, 1981 and 1991

| | Employment by industrial sector | | | | Change in employment by industrial sector 1981–91 | | |
| | 1981 | | 1991 | | | | |
	Number	%	Number	%	Number	% change	ppc*
Primary industries	57,275	1.6	41,364	1.3	−15,911	−27.8	−0.3
Manufacturing industry	683,951	19.2	358,848	11	−325,103	−47.5	−8.2
Construction	161,407	4.5	118,367	3.6	−43,040	−26.7	−0.9
Distribution, hotels and catering: repairs	686,598	19.3	645,955	19.8	−40,643	−5.9	0.5
Transport and communications	368,288	10.3	307,682	9.5	−60,606	−16.5	−0.8
Banking, finance, insurance, business services	565,876	15.9	733,513	22.5	167,637	29.6	6.6
Other services	1,034,526	29.1	1,049,015	32.2	14,489	1.4	3.1
Total services	2,655,288	74.6	2,736,165	84.1	80,877	3	9.5
Total employment	3,560,688	100	3,254,744	100	−305,944	−8.6	0

Sources: Censuses of Population, 1981 and 1991.

* ppc = percentage point change.

employment declined by 306,000 jobs (8.6 per cent). Manufacturing was by far the biggest loser with a decline of some 325,000 jobs: almost 50 per cent in a single decade (−47 per cent). Construction and primary industries also showed a sharp decline, losing a quarter of total jobs (−26.7 per cent and −27.8 per cent respectively). Transport and communications also declined by 16 per cent. On the positive side, two sectors showed gains: banking, finance, insurance and business services saw growth of almost 30 per cent (167,000 jobs) and other services increased by just 1.4 per cent. Overall, total losses of almost half a million jobs (485,000) were partly offset by gains in banking and finance. In proportionate terms, manufacturing industry continued its long decline from 19.2 per cent to 11 per cent of total employment, while the service sector expanded from 74.6 per cent to 84.1 per cent and banking and finance grew from 15.9 per cent to 22.5 per cent of employment.

Most routine manufacturing has either closed or been decentralised out of London or out of Britain. In the process, London lost virtually all its major manufacturing and materials-processing employers, including Hoover, Sainsbury's, Tate and Lyle and most of the printing industry which decentralised out of central London to cheaper sites in the outer South East and beyond. The furniture trade has largely gone, as has instrument engineering. All that is left are the light engineering, printing and clothing industries. The reasons for the decline are complex but need to be seen as part of the wider deindustrialisation of most major British cities during the last thirty years. The explanations which have been put forward include outmoded and cramped premises, outdated production facilities, planning constraints, high unionisation and high labour costs relative to those found in small towns, rural areas or in the less-developed world (Fothergill *et al.*, 1988; Martin and Rowthorn, 1986; Martin, 1988; Dennis, 1978; Danson *et al.*, 1980; Massey and Meegan, 1982).

Given the decline in total employment in London which accompanied the decline in manufacturing employment (total employment fell from 4.45 million in 1961 to 3.56 million in 1981, and 3.25 million in 1991), some took the view that London's economy appeared to be slipping away rapidly. Crampton and Evans (1992) suggested that 'the economy of London appears to have reached its apogee in the early 1960s and to have generally declined since that time' (p. 260). With the advantage of hindsight, this has proved incorrect, as the decline of manufacturing industry seems to have stabilised in the second half of the 1990s and the service industries grew strongly. What London may have experienced between 1961 and 1991 was a period of transition from an industrial to a post-industrial economy, compounded by major manufacturing job losses induced by the two oil price crises during the 1970s and inflation. Nonetheless, the decline in overall employment during the 1960s, 1970s and 1980s was puzzling as Buck, Gordon and Young note that London's industrial structure was favourable to growth by virtue of its over-representation in the nationally faster-growing service sector:

London should have benefited from the fact that it lacked agriculture and mining, and that the manufacturing sector with declining employment nationally, was increasingly under-represented in London, whereas services, which showed consistent gains nationally, were increasingly over represented . . . London had relatively more employment in industries with faster rates of national employment growth . . . If each industry in London had experienced the same rate of employment change as in the country as a whole, there should have been substantial increases in employment averaging 30,000 jobs per year, during the 1960s and through to the aftermath of the first oil price rises in 1974. Even in the recessions since 1974, only modest losses would have been expected in London, because the national decline was concentrated in manufacturing industries which were by then of less importance in London.

(Buck *et al.*, 1986, p. 66)

The explanation adopted by Buck *et al.* is that the decentralisation of manufacturing employed occurred not as a result of the direct out-movement of firms and plants, but because the differential birth and death rate of firms in London, particularly in Inner London, was adverse compared to other cities. Crampton and Evans (1992) argue that this may well have come about as a result of government policy. They point to five factors: first, the operation of the Location of Offices Bureau and the system of office development permits which served to discourage new office development in London; second, the operation of regional policy towards manufacturing industry, in particular the system of Industrial Development Certificates; third, the lack of a policy for and towards London. They argued that government policy towards London has been marked by indecisiveness such as over the high-speed rail link to the Channel Tunnel. Fourth, they point to the fact that since the abolition of the Greater London Council in 1986, London had no overall political authority until the formation of the Greater London assembly and the election of a mayor in 2000. Finally, they suggest the British town planning system, with its tight planning controls, may have pushed up land prices and choked off growth in London. Massey and Meegan (1982) are clear, however, that the causes of manufacturing job losses in London and the other major cities can be laid at the door of companies keen to maximise profits and cut the least profitable plants in a downturn. These tend to be the oldest ones, with higher production and labour costs.

Employment trends in the 1990s

Significantly, employment in London has grown strongly from 1993, the low point of the early 1990s recession, and by 1998 it was back to the levels of 1981. There was a strong growth in business services and other services and manufacturing employment stabilised. The manufacturing sector

Table 2.2 Employment change in London during the 1980s and 1990s

	Job change (1993–8)	Job change (1983–8)
Finance	25,000	58,000
Business services	210,000	126,000
Hotels and restaurants	56,000	16,000
Manufacturing	0	−142,000
All other	179,000	34,000
Total job change	470,000	92,000
Percentage change	15.2	2.7

Source: Banks and Scanlon (2000) from Annual Employment Survey.

employed some 280,000 (8 per cent) in 1998 compared to 282,000 in 1993. But Banks and Scanlon (2000) note that over 50 per cent of those employed in the manufacturing sector in London in fact work in the head offices of manufacturing companies who produce elsewhere. London's manufacturing base is now primarily one of command and control head office functions rather than direct production. As a consequence of these changes, while employment grew in the second half of the 1980s by 92,000, Table 2.2 shows that it grew by 470,000 or 15.2 per cent during the boom years of the 1990s. Banks and Scanlon comment that:

> Throughout the 1980s sharp losses in London's manufacturing industries kept overall job levels from rising above the 1980–81 threshold despite the strong growth which took place in finance and business services during the decade. By contrast, once past the recession years, manufacturing losses have stabilised in the 1990s while the strong employment gains in business services, the hospitality sector and in finance highlight the major driving forces in the London economy.
>
> (Banks and Scanlon, 2000, p. 40)

Although job growth in London's financial sector in the 1990s was lower than in the previous decade, as a result of the job losses which have taken place in retail banking, business services and the financial sector continued to be the main driving force in the London economy in the 1990s, with business services alone adding 200,000 new jobs (Banks and Scanlon, 2000). By 1999, finance and business services accounted for 32 per cent of employment, more than any other sector and four times that of the manufacturing industry. By comparison, finance and business services accounted for just 19 per cent of jobs in the UK as a whole, almost the same level as manufacturing. Employment growth in hotels and restaurants in London from 1993 to 1998 also increased by over 30 per cent, which is a reflection of the strength of tourism and consumer spending in the London economy during

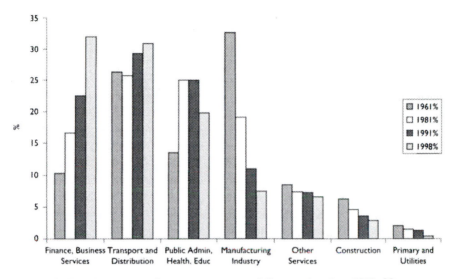

Figure 2.1 The changing employment structure of Greater London, 1961–98.
Sources: Derived from the Censuses of Population, 1961–91 and OWS.

Table 2.3 Employment by industry, 1999 (%)

	London	UK
Finance and business services	32.0	18.7
Distribution, hotels and catering	22.3	22.7
Education, social work and health	14.0	18.2
Transport, communications	8.6	6.0
Manufacturing	7.5	16.7
All other	15.6	17.7
Total	100	100

Source: *Focus on London* (2000), ONS.

this period (Church and Frost, 2002). It is impossible to disaggregate employment in tourism from other areas precisely, but Bull and Church (1996) estimate the tourist-related employment in hotels and restaurants accounted for 9 per cent of the total workforce in 1991 and employed 230,000 people in 1996.

Figure 2.1 shows the changing employment structure of Greater London from 1961 to 1998. Though the employment categories do not match precisely with those in Tables 2.1 and 2.3, it shows that over a forty-year period London's employment structure has been dramatically transformed from one where manufacturing was the largest sector to one where financial and business services are now collectively the largest sector. In 1961 manufacturing

Table 2.4 The regional concentration of employment in the cultural industries sector

Region	CIS % of employment		Location quotients	
	1984	1991	1984	1991
Rest of South East	5.7	5	1.2	1.1
East Anglia	4.3	4.4	0.9	1
London	7.7	7.1	1.6	1.6
South West	4.3	4.4	0.9	1
West Midlands	3.1	3.2	0.7	0.7
East Midlands	3.2	3.7	0.7	0.8
Yorks and Humber	3.4	3.7	0.7	0.8
North West	4	3.7	0.8	0.8
Wales	3.8	4.2	0.8	0.9
Scotland	3.9	3.6	0.8	0.8

Source: Pratt (1997).

industry accounted for 32 per cent of employees, and financial and business services for 10 per cent; by 1999 the proportions were reversed. There is another important sector in London's economy – that of the cultural industries. Although these are not disaggregated in Figure 2.1, Pratt (1997) has shown that in 1991 this sector accounted for 7.1 per cent of total employment in London (see Table 2.4), far higher than in any of the other regions. The London location quotient for the cultural industries sector (CIS) in 1991 was 1.6 (an index of 1 is average) which shows it is over-represented in the capital compared to other regions of Britain (see also Greater London Authority, 2002).

The changes in employment in different sectors parallel the shifts in gross domestic product or GDP. Looking at output measured in terms of real GDP by industry group in Greater London from 1981 to 1991, Graham and Spence (1995) show that in 1984 total services accounted for three-quarters (74 per cent) of total output. By 1991 this had risen to 78.6 per cent while manufacturing output fell from 17.9 per cent to 13.7 per cent. In percentage terms, manufacturing output decreased by 2 per cent while output in services rose by more than a third (36 per cent). The most dramatic increase, however, was in banking, finance, insurance and business services which rose by a remarkable 71 per cent. In absolute terms, the increase in output in this sector was greater than that of all other sectors combined and its share of total output rose from 22 per cent to 29 per cent. Greater London also saw an 11 per cent increase in the number of employment units from 1987 to 1991, but again the increase in the banking and allied services sector (28 per cent) was far greater than in any other sector. By 1996, GDP in financial and business services accounted for a remarkable 38.6 per cent of the London total compared to just 10.4 per cent in manufacturing industry (ONS, 1999).

The role of financial services in London's economy

Financial markets have been important in London from the early nineteenth century when it took over as the leading international financial centre from Amsterdam which had been occupied by the French in 1795. It retained its international dominance until the First World War when New York took over the leadership role. The City declined during the inter-war years, partly due to Britain's economic problems and subsequent regulation of foreign transactions. The Stock Exchange also adopted rather restrictive attitudes, including prohibition of options trading (Hyde and Dilnot, 2000). It was not until the 1960s and the development of the Eurobond market that London began to really take off again (Kynaston, 2001; Mitchie, 1992). Then, in the 1970s, the deregulation of the financial markets, the break up of the Bretton Woods system of fixed exchange rates set up post-war, the abolition of exchange controls in 1979 and the creation of a global financial marketplace enabled London to regain much of its previous importance (Lee and Schmit-Marwede, 1993; Coakley, 1992; Martin, 1994). Kynaston (2001) suggests that 'London is now a far more international centre than New York or Tokyo, whose financial sectors in large part live off servicing their domestic economies'. This is reflected in the fact that London now accounts for 58 per cent of the global foreign equities market and 70 per cent of the Eurobond market, as well as 32 per cent of world foreign exchange trading and 19 per cent of international bank lending. The City of London now accounts for $330 billion each day on the London International Financial Futures Exchange (LIFFE) and $637 billion a day in foreign exchange trading (City of London, 2000).

Table 2.5 shows the relative size of the world's twelve largest stock exchanges in 1999. It is apparent that, whether measured in terms of total

Table 2.5 World's largest stock exchanges by value and turnover, 1999

	Turnover 1999 ($bn)	Domestic value 1999 ($bn)	Number of listed companies
Nasdaq	9,986	5,036	4,829
New York	8,655	11,160	3,025
London	3,261	2,868	2,791
Tokyo	1,761	4,325	1,937
Germany	1,431	1,389	1,128
Euronext	1,301	2,405	2,038
Korea	735	296	725
Paris	709	1,437	1,149
Madrid	678	418	727
Switzerland	514	657	412
Italy	493	706	270
Taiwan	450	362	462

Source: London Stock Exchange, *International Comparisons*, 1999.

value or turnover or the number of listed companies, the New York stock exchange is far and away the largest in the world ($11,160 bn by value and $8,655 bn turnover). If the high-tech Nasdaq exchange is added it has a commanding lead. London ranks third after Tokyo with $2,868 bn by domestic value and second by turnover ($3,261 bn), ahead of Tokyo ($1,761 bn). The Paris–Amsterdam Euronext exchange comes in fifth by value and sixth by turnover.

The financial services sectors of London and New York are primarily concerned with wholesale rather than retail financial services – trading on foreign exhange, equity and bond markets, international banking and corporate finance, mergers and acquisitions, commercial insurance, fund management, and so on (Hyde and Dilnot, 2000). These activities have grown rapidly in the past two decades as a result of the growth of world trade, the internationalisation of investment, development of new products and services and the shift towards privatisation. The two cities have a leading role in almost every main financial market, including cross-border bank lending, foreign exchange dealing and foreign equities trading, fund management and mergers and acquisitions. London also performs a key role in international insurance with Lloyd's of London, and UK insurance companies took £14 bn gross premium insurance income in 2000.

The growth in employment in the financial sector in the 1990s has been relatively small because of the large number of staff lost in domestic retail banking during the period. Nonetheless, the City retains a leading role in wholesale banking and it has a leading role in international banking. It is estimated that there were 479 foreign banks in London in 1997 employing a total of 65,000 people: a marked increase from the 73 banks in 1960, and the number grew to 550 in the late 1990s. The recent fall is a result of a wave of mergers in the banking industry and the withdrawal of some Asian financial institutions as a result of the Asian financial crisis in the late 1990s (Clark and Wojcik, 2001; Kamo, 2000).

London is currently the main centre for both the issuance and secondary trading in the international bond market; that is: 'bonds denominated in a currency other than that of the country of issue, intended for international rather than domestic investors' (Hyde and Dilnot, 2000, p. 7). The Bank of England estimates that in late 1999 London's share of the Eurobond market was 60 per cent and that London also had 70 per cent of secondary trading in international bonds. In 1998 the London Stock Exchange accounted for 65 per cent of global foreign equity turnover and, until mid-1998 when German competition hotted up,

> LIFFE was unrivalled as the largest futures exchange in the world outside Chicago. Though Eurex is now the largest derivatives exchange, measured according to the volume of turnover, LIFFE remains the largest derivatives exchange according to the value of interest rate and equity index

Table 2.6 Identified net overseas receipts of UK financial institutions (£bn)

	1980	1986	1993	2001
Banking	0.1	4.1	5.6	3.2
Insurance	1.1	4.2	4.1	3.8
Fund managers	0.2	1	3.3	0.6
Commodities	0.3	0.4	0.6	—
Shipping	0.2	0.2	0.3	0.4
Security brokers, money brokers	0.2	1	3.6	3.9
Total	2.1	10.9	17.5	13.2

Source: British Invisibles, International Financial Services, London (2002).

> derivatives and contracts ... LIFFE dominates the market in Euro-
> denominated short-term interest rate contract market with a share of
> over 90 per cent.
>
> (Hyde and Dilnot, 2000, p. 7)

London is also home to a number of smaller securities and derivatives markets. The London Metal Exchange (LME) is the world's leading base metals exchange with an approximate 95 per cent share of the exchange traded market totalling $2,500 billion in 1997. The International Petroleum Exchange is Europe's leading energy futures and options exchange, with approximately 33 per cent share of the world oil market (Hyde and Dilnot, 2000). The Baltic Exchange is also the world's leading centre for ship charter-ing. Collectively, UK financial institutions, most though not all of which are based in London, have seen net overseas receipts increase from £2.1 bn in 1980 to £10.9 bn in 1986, £17.5 bn in 1993, and £13.2 bn in 2001 (see Table 2.6). This is a major contribution to Britain's overseas earnings.

Financial services employment is, however, quite strongly cyclical depend-ing on the strength of the global economy. Thus, although employment grew strongly during the mid-1980s in the run up to the 'Big Bang' as all the major banks recruited strongly for the projected growth of new business, there was a round of major job losses in the late 1980s and early 1990s as rationalisation took hold. Similarly, while there has been considerable job growth in merchant banking, investment banking and equity sales and research during the boom years of the 1990s, the sharp downturn in activity in 2001 is seeing major job losses as the financial institutions attempt to match staffing levels to the greatly reduced business levels. As staff costs are the greatest expense, usually running at over 50 per cent of gross earnings, a turndown in business volumes and profits necessarily means big staff cuts. It is estimated that 30,000 jobs may have been lost in financial services in the City of London in the second half of 2001 and the first half of 2002 (Hughes, 2001a; Pretzlik and Saigol, 2001). Hughes (2001a) quotes one city banker as saying: 'the Square Mile was staffed up for exceptional times. At a guess, I'd say the City is only half way through getting back to the right size.' This suggests that, like the late 1980s, the second half of the 1990s may have been an abnormal boom period in terms of activity, and that financial and related business service employment may decline in the early years of the twenty-first century.

The growing internationalisation of the City

The drive to increasing consolidation has also been accompanied by a rapid degree of internationalisation of the City. In recent years, a number of financial institutions have announced a move to London from other centres. The process began in the late 1980s with the arrival of a number of major American financial institutions such as Merrill Lynch, Goldman Sachs, JP Morgan, Credit Suisse First Boston keen to take advantage of 'Big Bang' and the deregulation of the City. This was followed in the early 1990s when several major City merchant banks, including Kleinwort Benson and Warburgs, were taken over by foreign institutions keen to establish a foothold in London. With the collapse of Barings, and its subsequent takeover by the Dutch ING Bank, the City has become dominated by overseas institutions (Auger, 2000) (see Table 2.7). The only remaining independent City merchant banks are Rothschilds, Lazards and Cazenove's, and the latter recently announced its intention to float in order to increase its capital base.

In the mid-1990s Dresdner Bank announced that it was moving its investment banking functions to London, and it has been followed by a number of major German banks including West Deutsche Landsbank. At the time this

Table 2.7 Major takeovers/sales of City financial institutions to overseas bidders

Bank	Purchases	Date	Price (£m)
Deutsche Bank	Morgan Grenfell	Nov 89	950
ING Bank	Barings Bank	Apr 95	660
			(nominal £1 plus debts)
UBS	SG Warburg	May 95	860
Dresdner Bank	Kleinwort Benson	June 95	977
Merrill Lynch	Smith New Court	July 95	511
Deutsche Bank	Nat West Equity	Dec 97	50
Société Générale	Hambros Banking	Dec 97	330
CCF	Charterhouse	Dec 97	240
Investec Group	Guinness Mahon	Apr 98	50
Deutsche Bank	BT Alex Brown	Jun 99	1,000
Chase Manhattan	Robert Fleming	Apr 00	4,880
Citigroup	Schroders	Jan 00	1,350

Source: Derived from Merrell (2000).

generated considerable debate in the national press, with the clear consensus being that, rather than posing a threat, this was a vote of confidence in the future of the city (Kaletsky, 1995; Rogers, 1996; Wilcock and Eisenhammer, 1994; Bowley, 1997; Gapper and Denton, 1996).

The process of consolidation has continued apace. In June 2001, Sony announced they were establishing their global treasury operation in London (Merrell, 2001b), Cantor Fitzgerald, the leading US wholesale Treasury Bond broker, announced they were closing down operations in Frankfurt and Paris and shifting them to London in late 2001 (Hughes, 2001b) and the *Financial Times* (Major, 2001) suggested that Deutsche Bank, Germany's largest bank, was contemplating moving its headquarters from Frankfurt to London given the growing international dimension to its business.

The City under threat?

London's position as an international financial centre is currently not as secure, and in future may not prove as enduring, as most commentators assume. It still remains the leading centre for Eurobond, international equities and foreign exchange business, but in each of these markets it is under threat from and likely to be overtaken by international competitors like New York and Tokyo.

(Coakley, 1992, p. 70)

Thus far, this has not happened. The implosion of the Japanese economy since the end of the bubble economy in 1992 and the collapse of the Japanese equities market has meant that Tokyo has temporarily ceased to be a major contender. A number of its major international banks have scaled back their international operations in the face of the continuing Japanese financial crisis and internal liquidity problems (Murphy, 2000) and Tokyo has slipped back down the league table of international financial centres. But, simultaneously, London has faced growing competition from European financial centres, notably Frankfurt, in a determined bid to challenge London's domination of the European financial markets. There has been fierce jockeying for position and there have been various attempts by the Amsterdam, Frankfurt and Paris bourses to overturn London's leading position and it was thought that the UK's refusal to join the Euro might undermine its position (Kaletsky, 1999; Brierley, 1998). This has not happened, however, and in fact London's dominance appears to have been strengthened in several key areas. In one area, however – that of German Bund trading – London lost its dominance very quickly when it maintained its open outcry trading in the face of the introduction of cheaper and faster electronic trading in Frankfurt. London's share of the bund market fell from 70 per cent to 4 per cent in a year (Harrison, 2000). Although LIFFE has now moved to electronic trading and abandoned pit trading this was a painful lesson.

In May 2000 the London and Frankfurt exchanges announced that they were to merge (Garfield, 2000b). The combined company was to be called iX and was to be based in London but with substantial operations in Frankfurt, and agreement had been reached for the Milan and Madrid exchanges to join later. The chairman of the London Stock Exchange made it clear that the proposed merger was the first step in a wider consolidation process to create an integrated Pan-European market. But in mid-2000 the merger was thrown into disarray by a hostile bid for the London exchange by OM Gruppen, owner of the Swedish stock market (Garfield, 2000b). When the bid failed, the merger collapsed with opposition from the British and German exchanges, both of whom feared they were giving too much away (Baker, 2000). Major consolidation is likely to take place, though, and in 1997 the Paris, Amsterdam and Brussels exchanges formed an organisation called Euronext to counter the London–Frankfurt tie up.

In September 2001, LIFFE put itself up for sale and found three potential bidders, the LSE, Deutsche Borse and Euronext. There was a fierce bidding competition to secure LIFFE (Boland and Pretzlik, 2001), but at the end of October LIFFE announced that it had accepted a £555 million all-cash bid from Euronext (Moore, 2001; Hughes, 2001c).

This came as a blow to the LSE who need to consolidate. Clara Furse, chief executive of the LSE, stated: 'We put forward a bold proposition for the creation of a combined business which would have created the first single technology platform for securities and derivatives. We are surprised that Liffe did not share our ambitious vision.' But the chairman of LIFFE rejected criticism that he had sold out to a foreign takeover, saying that 'We've never been a British business, we have always been internationally owned. This is good for London.' The head of Euronext said that the tie-up brought together LIFFE's strengths in fixed-income derivatives with Euronext's in equity derivatives and he stated: 'Every major financial institution has a presence in London and this will give us that presence.' Euronext are planning to use LIFFE to built a global derivatives platform based in London. What is clear is that Euronext were very keen to get a base in London, otherwise they faced being marginalised (Boland and Pretzlik, 2001). This decision was a blow to the LSE, but it is in discussions with a variety of other exchanges, including Euronext, Deutsche Borse and Nasdaq (Hughes, 2002). With international stock exchanges, size and market share are clearly crucial to retaining competitive dominance and more mergers seem likely.

Other business services

As indicated earlier, business services are now an extremely important element in the economy and labour force of global cities and London is no exception to this trend.

Over the period 1981–98 employment in real estate, renting and business activities in Greater London increased by 376,000, or 92 per cent, from 408,000 (11.6 per cent of total employment) to 784,000 (22.1 per cent). From just one in ten jobs in Greater London in 1981, the business and professional services sector accounted for one in five jobs by 1998 (Daniels, 2000). This is a major increase and highlights the importance of this sector for the London economy. It eclipsed the growth shown by the financial sector, which grew from 255,000 (7.2 per cent) in 1981 to 321,000 in 1988, then fell to 275,000 in 1993 in the aftermath of the early 1990s recession and rose to 300,000 (8.4 per cent) in 1998. While a substantial share of this sector's growth has been in areas such as residential estate agencies and non-central business services, it is nonetheless the major area of growth, far exceeding every other. As Daniels (2000) points out, business and professional services have consistently performed better than other areas of the London economy. In the period 1988–93 employment in London fell overall by 13 per cent while Business and Professional Services (BPS) grew by 1.9 per cent. It remains to be seen, however, to what extent business services are resilient in the face of the 2001 economic downturn which has hit advertising and consultancy revenues hard. Companies such as Reuters (provision of financial data and dealing screens), Pearson Longman (press and educational publishing), to give a couple of examples, have experienced sharp falls in revenue and are cutting jobs, along with the software houses which sell to the financial and business service sector.

Looking at the changing structure of Central London's employment in the period from 1971 to 1989, Frost and Spence suggest that Central London's evolution has increasingly been driven by its developing global functions rather than by its role in the domestic economy. They state that:

> The evidence of increasing specialization towards finance and business services is clear and simple. By 1989, Central London's economy was dominated by these activities (38 percent). No other sector of employment approached their size. All employment in public administration, education, health and other mixed consumer services only accounted for about 25 per cent of total employment, while all forms of distribution, hotel and catering work amounted to only 17 per cent. These figures compare to a figure of 11 per cent in finance and business services in the Rest of the SE excluding Greater London and 10.8 per cent in Great Britain as a whole. Central London thus had 3.5 times the percentage employment in this sector as the RSE.
>
> (Frost and Spence, 1993, p. 555)

Frost and Spence concluded that:

The role which Central London plays in the national economy changed substantially over the last twenty years. It has become a centre with a high level of specialization in finance and business services which, for much of the last decade, have been the sole focus of employment growth while decline has taken place in other forms of activity. Such development corresponds well to the notion of Central London evolving as a specialized centre for the provision of internationally orientated business services.

(Frost and Spence, 1993, p. 556)

These trends have increased during the 1990s. Daniels estimated that by 1998 almost one in four of all BPS jobs in Great Britain were located in Greater London, and the rate of growth in London has exceeded the rate in the rest of Britain. Some segments of BPS (computer and related activities, market research and consultancy services, advertising, labour recruitment and miscellaneous business services) that have increased their employment two to three times the national BPS average of 35 per cent from 1993 to 1997 have also increased their share of employment in London (Table 2.8). Daniels states that:

It is of course London's position as one of the most diverse financial centres in the world that has stimulated the demand for Business and Professional Services. These services perform a critical supporting role for national and international clients with operations in London, as well as directly becoming more engaged with international markets. The critical mass of demand is further consolidated by the fact that 24 per cent of Europe's largest companies (by market capitalisation) and more than 60 per cent of the Fortune Global 500 companies have offices in London, shares that are not rivalled by any other European city.

(Daniels, 2000, p. 41)

Looking at specific sectors, Daniels notes that there has been a growing concentration of leading European advertising agencies in London since the late 1980s as a result of a trend towards increasing sectoral concentration and the concentration of advertising expenditure in the hands of leading agencies. A series of mergers and acquisitions has led to the emergence of four large international groups in the EU, two of which are run from London. Where computer services are concerned, Daniels notes that proximity to the consumer is essential and London is the principal location in the UK for the headquarters of the subsidiaries of most of the leading international computer systems and software firms. Recruitment services are also very concentrated, with over 500 firms in London providing specialist services from the recruitment of temporary staff for routine tasks to head-hunting key personnel for leading companies.

Table 2.8 Employment in other business services, Greater London, 1993–7

Activity	1993	1997	1993–7	1993 (%)	1997 (%)	1993–7 (%)
Renting of machinery/equipment	15,361	18,289	2,928	2.99	2.64	19.06
Computer and related activities	40,921	79,735	38,814	7.95	11.51	94.85
Research and Development	14,655	12,267	–2,388	2.85	1.77	–16.29
Other business activities	443,536	582,329	138,793	86.21	84.07	31.29
Legal activities	54,464	60,029	5,565	10.59	8.67	10.22
Accounting, auditing, taxation	49,191	37,337	–11,854	9.56	5.39	–24.1
Market research/consultancy	9,514	16,373	6,859	1.85	2.36	72.09
7414*	40,796	48,659	7,863	7.93	7.03	19.27
Management holding services	6,977	17,804	10,827	1.36	2.57	155.18
Architectural and engineering services and technical consulting	56,982	55,483	–1,499	11.08	8.01	–2.63
Advertising	21,613	38,613	17,000	4.2	5.57	78.66
Labour recruitment/personnel	56,081	116,559	60,478	10.9	16.83	107.84
Investigation and security	20,364	24,838	4,474	3.96	3.59	21.97
Industrial cleaning	92,752	74,821	–17,931	18.03	10.8	–19.33
Miscellaneous business services	4,756	7,137	2,381	0.92	1.03	50.06
7482*	1,818	1,006	–812	0.35	0.15	–44.66
7483*	3,697	6,460	2,763	0.72	0.93	74.74
7484*	22,672	73,722	51,050	4.41	10.64	225.17
Total Greater London	514,473	692,610	178,137	100	100	34.63

Source: Daniels (2000), based on data supplied by NIERC, September 1999.

* Indicates category of business service.

Implications for the future

The focus on finance and business services and the cultural industries in the above account could be argued to paint an unduly rosy picture of the employment situation in London. It is therefore important to look at the implications of the very sharp fall in manufacturing employment in the two recessions of the early 1980s and early 1990s.

Unemployment in London has fallen sharply from its earlier peaks in the mid-1980s and the early 1990s recessions. In 1981 claimant count unemployment was 225,000. This rose to 396,000 (9.3 per cent) in 1986. It then fell to a low of 202,000 (4.8 per cent) in 1990 before rising to a second peak of 471,000 (11.6 per cent) in 1993. Since then it has fallen steadily to 204,000 (4.8 per cent) in 1999. However, claimant count figures under-represent the true level of unemployment as not all unemployed persons are eligible for benefit. On the ILO definition, unemployment reached 13.7 per cent of London residents in 1993 and has since fallen by almost half to 7.5 per cent in 1999. But unemployment remains much higher in Inner London, particularly inner east London, and the implications of this for social deprivation and exclusion are discussed in Chapter 8.

Although manufacturing is no longer a significant component of London's economy, it is by no means immune to the effects of economic recession. Indeed, the reliance of London on financial and business services and tourism means that when there is a downturn it may be affected. Buck and Gordon (2000) point out that unemployment in London has tended to rise relative to that in the rest of the South East (ROSE) over the last twenty-five years, particularly when the demand-deficient unemployment rises across the region.

The potential danger for London's economy is that with the decline of manufacturing industry and the dramatic growth of financial and business service employment it has become over-reliant on a particular sector of economic activity which is susceptible to sharp changes in global economic and financial conditions. Thus the second half of the 1990s saw the development of a major financial bubble in the American stock market (Brenner, 2000) with the Nasdaq rising from 1,500 to a peak of 5,500 in March 2000, before collapsing to 1,400 in 2003. While the USA may just have avoided a recession, the dotcom bubble has collapsed, the telecommunications industry is in deep difficulties, and the market for company flotations, new issues and mergers and acquisitions is deeply depressed. With market activity depressed, financial institutions are laying off staff at a rapid rate and it is estimated that there may be up to 30,000 jobs lost in the City (Hughes, 2001a). The impact of the cuts could be severe in terms of their knock-on impact on demand for business services, but this would not be the first time that this has happened. A similar wave of redundancies took place at the end of the 1980s when financial institutions found themselves overstaffed in the wake of the Big Bang. The financial and business services sector seems to be quite cyclical in terms of activity and employment, but in the long term the sector seems set to continue to grow in importance as financial flows and business services assume a greater importance in the leading cities of the global economy.

It is also important to put reliance on financial and business services into perspective. To pose a rhetorical question: if this sector is so problematic why is London engaged in fierce competition with Frankfurt and other European financial centres to maintain its dominance? The answer of course is that financial and business services contribute a disproportionate share of GDP and are seen as one of the leading edges of contemporary economic growth. Reliance on finance and business services carries risks, but so does reliance on routine manufacturing or on high-tech industry, as the current downturn in telecommunications illustrates. Similar arguments apply to tourism and the cultural industries. London is one of the world's leading tourist cities and tourism contributes some £8 bn per annum to London's economy, but the industry received a worldwide shock in the aftermath of the 11 September 2001 WTC disaster when people stopped flying. This hit London and New York very hard, but nobody is arguing that London should try to reduce the importance of the tourist industry. On the contrary, the head of the London Tourist Board resigned in April 2002, claiming that the slump in visitors was not being addressed strongly enough and that London risked losing its tourist world city status unless action was taken (Gray, 2002). The conclusion seems to be that there are very few, if any, economic safe havens in the modern world, but that it is better to be over-reliant on the growth sectors than the declining sectors. In this respect, London is in a very strong position. As Gordon (2001) notes: 'London represents the core of the economically strongest region in the UK (if not of the EU)' (p. 25). This is a fortunate position to be in, and, notwithstanding the major job losses in London's manufacturing sector, it is one that many of the declining industrial regions would willingly swap.

Chapter 3

The transformation of London's occupational structure and the rise of the new middle class

London is ceasing to be a city with a large, organised factory proletariat and is reverting towards a capital in which a huge, underemployed proletariat works to service a wealthy minority. London is slipping backwards in time and becoming a plebeian city, a restless mass no longer to be regimented and organised by union or by class-party. This is more like Paris in the 19th century than the working class London of Herbert Morrison 40 years ago.

(Ascherson, 1986)

The image conveyed by the term 'polarisation' (alias the 'rout of the middle classes') is, first, that both the top and bottom groups in the society of London are becoming larger, at the expense of the middle strata; second, that the extreme groups are, or will be, located in sharp juxtaposition to one another – on either side of the 'tracks'. Both these images are false. In fact, the upper and middle strata are becoming stronger, numerically, in Greater London, and the skilled manual group has remained stable, while the proportion of non-skilled workers has decreased.

(Glass, 1973, pp. 423–4)

Introduction: what is happening to the occupational class structure of global cities?

What is happening to the occupational class structure of London and to other global cities such as New York and Paris – and does it matter? The answers to these questions have kept social scientists, planners, social reformers, politicians and journalists busy for the last hundred years, if not longer. In particular, there is a recurring question of whether the social-class composition of such cities is shifting upwards or downwards. As the quotes of Ascherson and Glass indicate, there are markedly divergent views on this issue. Put simply, there are three principal interpretations. The first is that the class structure of London (and other world cities) is becoming increasingly *proletarianised*, with a large and growing population of low-skilled service workers. The second can be termed *professionalisation*, and argues that the class structure of London and other global cities has been marked by the growth of professional, managerial and technical occupational groups and a decline of manual workers. The third interpretation (*social polarisation*) can be seen as a combination of the previous two and argues that there is growth at both the top and the bottom ends of the occupational and incomes structure with decline in the middle. Marcuse nicely summarised the polarisation thesis as follows:

> The best image . . . is perhaps that of the egg and the hour glass: the population of the city is normally distributed like an egg, widest in the middle and tapering off at both ends; when it becomes polarised the middle is squeezed and the ends expand till it looks like an hour glass. The middle of the egg may be defined as intermediate social strata . . . Or if the polarisation is between rich and poor, the middle of the egg refers to the middle income group . . . The metaphor is not structural dividing lines, but of a continuum, whose distribution is becoming increasingly bi-modal.
>
> (Marcuse, 1989, p. 699)

The potential implications of these radically different views are profound. The first interpretation, that of proletarianisation, is arguably most problematic, suggesting a future of growing deprivation and poverty as the middle classes flee the central city for the safety and social desirability of the suburbs. This is the situation which has beset a number of major American cities such as Detroit (Beauregard, 1993; Deskins, 1996). While the second view, that of professionalisation, could be seen as beneficial in terms of skill levels, purchasing power, education, segregation and the like, it has major implications for access and affordability to the housing market. In particular, the growth of the middle class, and the associated gentrification in inner-city residential areas, could squeeze out the existing working-class population. The third view, that of social polarisation and social dualisation, has become the conventional wisdom in the 1990s, although there is often little precision in the way the term is used.

The implications of these interpretations are very different. On the one hand, Ruth Glass argued thirty-five years ago that London was not becoming a more polarised or a more proletarianised city but was instead becoming more middle class, and that the working classes are being gradually squeezed out through the operation of the job and housing markets. On the other hand, Neil Ascherson suggests, like Sassen, that while London is undoubtedly witnessing growth at the top, this is in turn causing growth at the bottom end with a growing army of casualised workers. The analysis which is put forward here of what is happening to the social structure of London and other global cities such as Paris and New York is far more closely aligned to that of Ruth Glass than other commentators. But before we discuss competing interpretations, and the evidence for what is happening in London today, it is valuable briefly to look back historically.

The nineteenth-century fear of the dangerous classes

Concerns about the changing social structure of cities and their effects are not new. Historically, the dominant nineteenth-century middle-class concern was that the class composition of cities was shifting downwards as a result of the rapid industrialisation, urbanisation and migration of unskilled labourers to cities (Engels, 1969; Glass, 1968). Such fears initially focused on the rapidly growing northern industrial towns during the first half of the nineteenth century but, as Stedman Jones (1971) pointed out, 'In the period after 1850, fears about the consequences of urban existence and industrial society centred increasingly on London. London, more than any other city, came to symbolise the problem of the "residuum"' (p. 12). These fears reflected the large size of London's casual labour market, the effects of migration and the huge size of the city itself. Victorian London was by far the largest city in the world, and during the course of the nineteenth century its population grew from just over one million to over six million, and the growth of the working classes in south London and the East End was the focus of much middle-class concern regarding the impact of urban 'degradation' and the so-called 'residuum'. As Stedman Jones comments:

> The location of the residuum in London was particularly unsettling if only because of the immense size of the city and its national importance . . . Vast tracts of working-class housing were left to themselves, virtually bereft of any contact with authority except in the form of the policeman or the bailiff. The poor districts became an immense terra incognita periodically mapped out by intrepid missionaries and explorers who catered to an insatiable middle-class demand for travellers' tales. These writers sometimes expressed apprehension about the large and anonymous proletarian areas of South London, but the most extensive and the most feared area was the East End.
>
> (Stedman Jones, 1971, p. 14)

The problem of class segregation in nineteenth-century London was accentuated by growing out-migration of the middle classes and skilled tradesmen from the central areas to the suburbs as a consequence of the influx of low-skilled migrants. The vicar of Clerkenwell stated in 1857 that 'The richer classes are continually moving to other localities and the poor are taking their place. Houses which were filled with tolerably well-to-do, are now let out in lodgings' (quoted in Green, 1986, p. 123), and the vicar of the parish of St George in the East pointed to the role of differential out-migration of the middle classes:

> Most of the better class of tradesmen have migrated to the suburbs – to Stratford, Bow, Hackney and elsewhere, and the difficulty of obtaining funds for the various charities is increasing each year, whilst every year the need for them is increased, for as the better class go the poor fill their place. Few will soon be left besides the poor, the poor-rate collector, the relieving officer, the policeman and the parson.
>
> (Green, 1986, p. 124)

These concerns were not confined to London. Similar accounts can be found for other major nineteenth-century cities such as Liverpool, Manchester, New York and Chicago which grew rapidly as a result of migration, often from abroad (Banfield, 1974; Ward, 1989). These fears faded away somewhat in the last two decades of the nineteenth century and the early years of the twentieth as the housing conditions of the working classes improved and they slowly began to suburbanise and join the middle-class exodus from the cities as a result of the introduction of working men's railway fares. When, says Glass (1968): 'Masterman in 1909 asked why London unlike other European capitals, had not produced any revolutionary programmes, he saw a partial answer, at least, "hidden in the little comfortable two-storied red and grey cottages . . . proclaiming with their cleanliness and tiny gardens and modest air of comfort a working population prosperous and content"' (p. 69). To the extent that Glass is correct in suggesting that the growth of suburbanisation helped simultaneously to improve working-class living conditions, to tame working-class radicalism, and to sooth middle-class fears, it is not suprising that the great expansion of pre-war and inter-war owner-occupied suburbia (Jackson, 1974) tended to be characterised by a lack of debate on social composition of the inner cities. Similarly, the post-war years were marked by considerable social optimism characteristically associated with the large-scale expansion of the New Towns and public and private house building.

The re-emergence of concern

It was not until the late 1960s and early 1970s, in the aftermath of the urban riots in the USA, that fears began to re-emerge regarding the social composition of London. The precise nature of the fears varied considerably, but the origins of the polarisation debate lay in the concern first voiced in the Greater London Development Plan (GLC, 1969) that London could be facing a similar situation to that experienced in the USA post-war, where the 'flight to the suburbs' of the middle classes and the concentration of the poor and ethnic minorities in the inner cities had led to an apparent polarisation of social classes. A related concern expressed by Wilmott (1969) and the South East Joint Planning Team (SEJPT, 1970) was that a process of bipolarisation was occurring whereby the inner areas were losing the intermediate social stratum of the population by out-migration, leaving a city of the very rich and the very poor. This phenomenon was termed the 'rout of the middle classes' by Glass (1973) who firmly rejected both possibilities in favour of the view that 'The real risk for London is that it might well be gentrified with a vengeance and be almost selectively reserved for selected higher-class strata' (p. 426). The gentrification debate is taken up in more detail in Chapter 7, which argues that what Glass foresaw has come to pass. The following section outlines the three major theses regarding social change in contemporary cities and evaluates their strengths and weaknesses before turning to an empirical examination of the changes in occupational class composition in London from the early 1960s onwards.

Proletarianisation versus professionalisation

The traditional Marxist view is that all capitalist societies are undergoing a process of *proletarianisation* as changes in the labour process designed to increase profitability lead to greater mechanisation and automation of production, deskilling and growing immiseration of the working classes. This is consistent with Marx's view that there would be a gradual withering away of the petite bourgeoisie and other intermediate class fractions and the emergence of a sharp divide between the ruling class and an expanded, and immiserated, working class. As Butler points out:

> the so-called pauperisation thesis (that as capital accumulates, the proportion of value available for wages by necessity lessens) was central to Marxism . . . Marxists hung on to abstracted notions of the falling rate of profit and the growing rate of exploitation allowing them to argue that the working classes were becoming increasingly immiserated despite the torrent of consumer durables that invaded working class homes.
>
> (Butler, 1999, p. 88)

Braverman (1974) elaborated this thesis for white-collar work, arguing that despite the growth of white-collar work, changes in the organisation of work (such as telephone call centres) are also leading to systematic deskilling and proletarianisation of white-collar workers. Generally speaking, this thesis has been applied to capitalist societies as a whole, rather than to urban areas in particular, although an extensive literature has developed on the emergence of a so-called urban underclass (Banfield, 1974; Murray, 1984; Wilson, 1987, 1996; Mingione, 1996; Hamnett, 2000).

The second, and contrary, view, often strongly associated with the work of Daniel Bell (1973), is that capitalist societies are undergoing a change from industrialism to post-industrialism. This fundamental shift is associated with the shift from a manufacturing- to a services-based economy, the declining importance of manufacturing employment and the growth of a quaternary (information processing) economy. According to Bell, this shift is linked to the growing importance of technical and professional knowledge and education, a decline of manual work and the growth of non-manual employment, particularly in professional, technical and managerial occupations. The post-industrial thesis was developed by David Ley (1980) in pioneering work on the changing class and political structure of post-industrial cities. He argued that the shift from a goods-producing to services-producing economy, the declining role of unskilled labour in the production process and the growing importance of technology in the factory and in offices, represents a major break with the nineteenth century. Ley saw this as linked to a major transformation of the labour force, with a decline in blue-collar workers and the growth of white-collar workers, particularly in the professional, managerial, technical and administrative occupations. He suggested that:

> we may see from this framework the appearance of a theoretically significant group of actors . . . [who] form a theoretical counterpoint to nineteenth century notions of capital and labor . . . a class in emergence . . . With a secure economic base they represent the present day counterparts of Veblen's leisure class, displaying the canons of good taste, intent upon the aesthetic. Their lifestyle is consumption and status orientated in pursuit of self-actualisation.
>
> (Ley, 1980, pp. 242–3)

This thesis, which is linked to the emergence of a growing new middle class based on technical and professional skills, has proved anathema to many Marxists (Walker and Greenberg, 1982; Smith, 1987) as it runs directly counter to Marx's views regarding the increasing polarisation of the class structure between a small ruling class who possess increasing control over the ownership of the means of production and a large and growing deskilled proletariat with no control over the means of production and nothing to sell but their labour power. Neil Smith (1987), for example, accepts that there is no doubt that 'Western capitalist economies have experienced a decline in the relative importance of manufacturing employment and a parallel increase in the importance of professional, administrative service, and managerial occupations', but he argues that it 'does not prove the emergence of a new middle class' (pp. 153–4).

The differences between post-industrial and Marxist proletarianisation theorists were outlined by Wright and Martin, who observed that:

Technological change is generally viewed by post-industrial theorists as increasing the cognitive complexity of work and the demand for experts while generally eliminating routinized, and especially, manual jobs. As a result, post-industrial theorists generally expect a process of gradual de-proletarianization characterized by an expansion in the semi-autonomous employee category and a decline in the working class. Marxists, however, generally postulate an inherent tendency in capitalism for technical change to destroy highly skilled, cognitively complex jobs and to reduce the autonomy and self-control of wage earners in the labor process. While technical change may also create new kinds of skilled positions, overall there should be a net tendency for working-class jobs to expand relatively to semi-autonomous or expert jobs. It would generally be predicted, therefore, that transformation of class structure should be characterized by a continuing process of increasing proletarianization.

(Wright and Martin, 1987, pp. 2–3)

There is little doubt that this has taken place in many areas of manufacturing industry as a result of automation and the decline of skilled work, and Wright and Singlemann (1982) tested these theses for the USA using data for 1960–70. They found there were simultaneously processes of proletarianisation and de-proletarianisation during the 1960s. But in subsequent work on the transformation of the American class structure from 1960 to 1980, Wright and Martin (1987) found that 'far from an acceleration of tendencies towards proletarianization observed in the 1960s, the pattern in the 1970s appears to be a direct reversal of these tendencies . . . The implication of these analyses is unmistakable: the results are more consistent with the post-industrial society thesis than with a traditional Marxist proletarianization thesis' (p. 18). They added that the results:

pose a real challenge to traditional Marxist assumptions about the trajectory of development of class structure of advanced capitalist societies in general, and about the process of increasing proletarianization in particular.

(Wright and Martin, 1987, p. 19)

As Wright is one of the leading Marxist class theorists, these conclusions cannot simply be dismissed by Marxists as anti-Marxist, and therefore wrong, and similar conclusions have been made in most major western countries. Myles (1988) found that in Canada 'the years from 1961 to 1981 were a period not only of continued but acceleration of new middle class jobs' (p. 250), and work by a variety of other class analysts points clearly in the same direction. Marshall and Rose (1985) examined the evidence for proletarianisation of the British class structure and found it very weak.

Criticisms of the professionalisation thesis

Much of this research is based on the analysis of official statistics on skill levels and occupational class, and the critics argue that there are major classificatory problems. Whereas the professional socio-economic groups, both self-employed and employees, are distinguished on the basis of work 'normally requiring qualification of University degree standard', and that of employers and managers on the basis that they employ, supervise or plan the work of others, the sole characteristic intermediate and junior non-manual workers have in common is the non-manual status of their work. In other respects they embody major differences in skill levels, responsibilities and incomes. As Braverman (1974) has pointed out, however, 'The traditional distinctions between "manual" and "white-collar" labor, which are so thoughtlessly and widely used in the literature on this subject, represent echoes of a past situation which has virtually ceased to have any meaning in the modern world of work' (p. 325). Braverman's now classic analysis of white-collar work in the twentieth century revolves around the fact that whilst there has been a massive increase in the size of the non-manual office and sales labour force during the course of the last eighty years (see Marsh, 1965), this growth has been associated with a significant downgrading of the nature of the work involved, skill levels and responsibility. Braverman argues that whilst the use of the term 'white-collar'

> tends to get its occupational flavor from the engineers, managers and professors at the top of the hierarchy . . . its impressive numerical masses are supplied by the millions of clerical workers, in much the same way that the stars of the opera company occupy the front of the stage, while the spear-carriers provide the massive chorus.
>
> (Braverman, 1974, p. 350)

The root of the problem, suggests Braverman, is that the clerical office workers of today are radically different from those of yesteryear. In the nineteenth century the office was the site of a small number of relatively highly paid and responsible clerks who carried out the major planning and supervisory roles for a much larger number of manual workers. Today, however, the deskilling and proletarianisation of routine office work has meant that the typical contemporary office worker is non-manual in name only. Braverman argues that:

> if one ascribes to the millions of present-day clerical workers the middle-class or semi-managerial functions of that tiny and long vanished clerical stratum of early capitalism, the results can only be a drastic misconception of modern society.
>
> (Braverman, 1974, p. 29)

Braverman provides evidence which shows that in the United States in 1971 'the median usual weekly wage for full-time clerical work was lower than in every type of so-called blue-collar work' (p. 297). Lockwood (1958) identified a similar pattern in Britain in the late 1950s: 'the average clerk is now very roughly on the same income level as the average manual worker, or perhaps even slightly below'. Thus, Braverman argues that '*the apparent trend to a large non-proletarian "middle-class" has resolved itself into the creation of a large proletariat in a new form*' (1974, p. 355; emphasis added).

A similar analysis has been made by Kumar (1978) who argues that whilst the post-war growth of the professional and technical group has been extremely impressive, it is rare that the census occupational statistics are critically interpreted. Professional and technical status has been assigned to a very diverse range of occupations leading to the growth of the 'semi-professions', but Kumar argues that:

> The clear evidence is that the vast bulk of expansion in these fields has been at their lower ends, on almost any scale we care to choose – qualifications, autonomy, pay. The growth is dominated by school teachers, nursing staff, junior and relatively unqualified social workers, and ancillary technical staff of various kinds. Their conditions of work have little in common with that of the traditional professional, and indeed shade off readily into the sphere of routine white-collar work.
>
> (Kumar, 1978, p. 216)

While Braverman and Kumar are undoubtedly correct regarding the growth of routine white-collar work and the upward creep in occupational classifications, the changes in occupational structure which have taken place nonetheless represent a very marked shift from the earlier dominance of the manual working class, and it is misleading and partial to suggest that the changes which have taken place are minimal or cosmetic. Halsey (2000) states that 'As a proportion of those in employment, manual workers fell from three-quarters to just under half between 1911 and 1981 and further to just under a third by 1991.' He notes of social mobility in Britain more generally that:

> By the end of the century millions of children of manual workers had risen into non-manual jobs and many thousands had become the graduate grandchildren of butchers, bakers and candlestick maker, following professional careers. The occupational, and therefore the class structure had shifted. In 1900 the vast majority of Britons were elementary school proletarians; by 1970 they were divided half and half between white and blue collar jobs. By 2000 the balance had been tipped decisively to form a white collar majority.
>
> (Halsey, 2000, p. 17)

A similar occupational class shift has characterised most western developed countries and there has also been a marked improvement in levels of educational qualifications, with a much higher proportion of the workforce possessing university degrees or other higher educational qualifications than in previous decades (Hamnett, 1994a). While the middle class has become less elitist and more variegated than in the past, to claim as Braverman does that most of its members are little more than white-collar proletarians in a new guise seems to be stretching things too far. Esping-Andersen (1993), in a study of trends in the changing class structure in six western countries, states that 'Virtually all research rejects Braverman's de-skilling and mass proletarianization thesis while pointing, instead, to a pervasive momentum of skill upgrading and professionalization' (p. 32).

The polarisation thesis

The third and most recent thesis, that of social polarisation, emerged in the USA in the 1980s and was associated with concern about what was termed the 'shrinking middle class' (Kuttner, 1983; Lawrence, 1984; Levy, 1987). It was argued that a combination of deindustrialisation and a shift to the service sector was leading to the loss of skilled manufacturing jobs and a polarisation of the occupational class structure between an elite of highly skilled professional and managerial workers at the top of the service sector and unskilled service workers at the bottom (Noyelle, 1983). Friedmann and Wolff (1982) linked development of social-class polarisation to the economic role and employment structure of world cities which are seen to be the command and control centres of the capitalist world economy. In particular, they pointed to the growing importance of financial and business services:

> A primary fact about world cities is the impact which shifts in the structure of their employment will have on the economy and social composition of their population. The dynamism of the world city economy results chiefly from the growth of a primary cluster of high-level business services which employs a large number of professionals – the transnational elite – and ancillary staffs of clerical personnel.
> (Friedmann and Wolff, 1982, p. 320)

They identified two other growing clusters of employment. The first of these serves the first and consists of real estate, construction activity, hotels, restaurants, luxury shops, entertainment, private security and domestic services. The second, overlapping the first, focuses on international tourism and Friedmann and Wolff (1982) suggest 'The growth of the first three clusters is taking place at the expense of manufacturing employment. Although a large cluster, its numbers are gradually declining as a proportion of all employment' (p. 320).

They also identify a government service cluster which is concerned with maintenance and reproduction of the world city, and the 'informal', 'floating' or 'street economy' which is particularly, though not exclusively, concentrated in the semi-peripheral cities of the Third World where it can be the largest cluster of employment. They argue that:

> Transnational elites are the dominant class in the world city, and the city is arranged to cater to their life styles and occupational necessities . . . The contrast with the third or so of the population who make up the permanent underclass in the world city could not be more striking . . . *The primary social fact about world city formation is the polarisation of its social class divisions.*
>
> (Friedmann and Wolff, 1982, p. 322; emphasis added)

Sassen's polarisation thesis

The polarisation thesis has subsequently been developed by Sassen (1984 [as Sassen-Koob], 1988, 1991) who argues that the changes in social structure are a direct result of the changes in the economic base. She argues that the evolving structure of economic activity in global cities, particularly the rapid growth of financial and business services and the sharp decline of manufacturing industry, has 'brought about changes in the organisation of work, reflected in a shift in the job supply and polarisation in the income and occupational distribution of workers' (1991, p. x). In addition, Sassen argues that:

> Two other developments in global cities have contributed to economic polarisation. One is the vast supply of low-wage jobs required by high-income gentrification in both its residential and commercial settings. Furthermore, there is a continuing need for low-wage industrial services, even in such sectors as finance and specialised services. A second development that has reached significant proportions is what I call the downgrading of the manufacturing sector, a process in which the share of unionised shops declines, and wages deteriorate while sweatshops and industrial homework proliferate.
>
> (Sassen, 1991, p. 9)

Sassen summarises her thesis as follows:

> New conditions of growth have contributed to elements of *a new class alignment in global cities*. The occupational structure of major growth industries characterised by the locational concentration of major growth

sectors in global cities in combination with the polarised occupational structure of these sectors has created and contributed to growth of a high-income stratum and a low-income stratum of workers. It has done so directly through the organization of work and occupational structure of major growth sectors and it has done so indirectly through the jobs needed to service the new high-income workers, both at work and at home as well as the needs of the expanded low-wage work force.

(Sassen, 1991, p. 13)

Dual cities?

Similarly, in their influential book *Dual City: Restructuring New York*, Mollenkopf and Castells (1991) posed the question of whether New York is a 'dual city', split into two distinct and separate parts, one of wealth and splendour, the other of poverty and oppression. They argued that, while New York is a very unequal city, contrary to popular impression and academic suggestion it is *not* a dual city:

Have the last two decades actually produced a New York composed of two separate and unequal cities as the popular image would have it. If one could answer this question by looking at trends in income inequality and poverty, the answer would clearly be yes. But the dual city metaphor . . . is flawed as an analytic approach. The 'two cities' of New York are not separate and distinct, but rather deeply intertwined products of the same underlying processes.

(Mollenkopf and Castells, 1991, p. 11)

And again, in their conclusion, they argue that:

the dual city is a useful ideological notion because it aims to denounce inequality, exploitation, and oppression in cities . . . But its underlying assumptions are rarely made explicit, because those who employ it tend to favor social critique over social theory. The political and emotional charge of a dualist approach and the failure to spell out its assumptions means that it cannot comprehend the complexity of urban social reality, which is certainly not reducible to a simple dichotomy.

(Mollenkopf and Castells, 1991, p. 405)

Yet, notwithstanding their conclusions that New York is not a dual city, the notions of dual cities and social polarisation have uncritically passed into conventional wisdom on the nature of urban social change. Mollenkopf and Castell's book is often referred to in the literature as if New York was a dual city (but see Marcuse, 1989). The trends are taken for granted and the validity of the thesis is assumed rather than examined.

The polarisation thesis: a critical assessment

The polarisation thesis developed by Friedmann and Wolff and Sassen is a fascinating and compelling one which links changes in the national and international division of labour, concentration of financial and business services, the decline of manufacturing industry and employment, and growth of a low-skill service sector in global cities, to changes in occupational and income distribution in these cities and to growing social segregation. It links together a number of observable trends into a set of coherent theoretical and empirical propositions. The question, however, is not how attractive the thesis is, but to what extent it is theoretically and empirically valid. It is argued that while parts of the polarisation thesis are undoubtedly correct, others are ambiguous and some are empirically invalid. As Fainstein *et al.* (1992) pointed out:

> The images of a dual or polarised city are seductive, they promise to encapsulate the outcome of a wide variety of complex processes in a single, neat and easily comprehensible phrase. Yet the hard evidence for such a sweeping and general conclusion regarding the outcome of economic restructuring and urban change is, at best, patchy and ambiguous. If the concept of 'dual' or polarising city is of any real utility, it can serve only as a hypothesis, the prelude to empirical analysis, rather than as conclusion which takes the existence of confirmatory evidence for granted.
>
> (1992, p. 13)

While Friedmann and Sassen are clearly correct regarding the declining importance of manufacturing industry and employment in global cities, and the growth of financial and business services and employment, the social polarisation thesis is problematic for a number of reasons. First, it is beset by definitional ambiguity, particularly regarding measurement of polarisation; second, it appears to contradict a large body of existing work on social-class change in advanced capitalist countries; and third it is unclear to what extent the thesis is over-dependent on the experience of high immigration cities such as Los Angeles and New York which may have a very specific labour market with a large pool of low-skilled immigrant workers (legal and illegal) who can readily be recruited into low-paid service jobs (Hamnett, 1994a, 1996). This is much less true of European cities, notwithstanding the growing proportion of minority groups, as a result of differences in the structure of the labour market and in the state regulation of labour markets and employment conditions. Although there is no doubt that low-wage immigrant workers are employed in the service sector, Europe has not seen the growth of low-skill service-sector jobs on anything like the same scale as the USA.

To take the problems in order, although there can be no doubt that Friedmann and Wolff (1982, p. 322) are right in claiming that 'the transnational elite is the dominant class in the world city', it is unclear what precisely they mean by 'the polarisation of social class divisions' other than that there are sharp and growing class divisions between different strata – which seems indisputable. Sassen argues that changes in the division of labour and the organisation of production in global cities are leading to a decline in middle-skill jobs and an increase in both high- and low-skilled jobs, and a concomitant increase in low- and high-wage jobs. What is unclear, however, is to what extent she sees these changes in absolute or relative terms. There is a major difference between asserting that these changes are creating large numbers of low-skilled and low-paid jobs, and changes which lead to a larger proportion of low-skilled and low-paid jobs but not necessarily larger numbers. Sassen seems to suggest that a process of absolute polarisation is occurring but, though this may be true in New York or Los Angeles, where a large migrant labour force is available, in most western capitalist countries the numbers and proportions of semi-skilled and unskilled jobs have been steadily shrinking for the last 30–40 years. The problem in London and key European cities may be not that there are too many low-skilled jobs but that there are not enough to provide work for those with few educational qualifications. A consequence of this is rising unemployment (Hamnett, 1994b, 1996; Burgers, 1996; Gordon, 1999). Indeed, Esping-Andersen (1993) points out that while the empirical studies of polarisation have focused on earnings rather than jobs, it is also possible to envisage a cleavage between labour market insiders (with high wages and job security) and a growing army of outsiders, including youth, long-term unemployed, discouraged workers and early retirees. If this is what is meant by polarisation it may indeed be happening, but it will not be shown up in the skill structure of the labour force, as Sassen suggests, as the key issue is those groups who are outside the economically active and employed. The fundamental problem with the proletarianisation thesis is the direct empirical one that it runs counter to almost all the available evidence on the changing class structure of advanced societies, which points to a strong upward shift in the occupational class structure and a sharp decline in the number and proportion of manual workers at both the national and urban level. Esping-Andersen concludes that

> there is little evidence to suggest strong polarisation. Everywhere, the trend favours the higher grade occupations such that the shape of the post-industrial occupational hierarchy is biased towards the top and the middle, rather than the bottom.
>
> (Esping-Andersen, 1993, p. 53)

On the other hand, critics such as Bruegel (1996) and Cox and Watt (2002) argue that much of the analysis of changes in occupational structure relies on aggregate census and other official data which is based on the economically active, and inevitably under-counts or misses those groups who work in the informal sector to supplement benefits or because of their immigration status. They argue that informal survival strategies are an important element of the urban economy which are not picked up by official data. Whilst this may indeed be happening, the problem is that we know little about it, as they themselves admit.

The changing occupational-class structure of London

The objective in this section is to examine the data on occupational-class change in London over the last forty years in order to assess empirically the polarisation thesis. One of the first studies to examine the changing social-class composition of London in the post-war period was undertaken by Wilmott and Young who looked at the period from 1951 to 1966. They concluded that:

> There was a general upgrading. Both Greater London and the Outer Metropolitan Area showed an increase in the proportion at the top and a decrease at the bottom, with the intermediate class staying much as it was. In this sense, the whole London Region became more middle class over the fifteen years.
>
> (Wilmott and Young, 1973, p. 200)

More recent evidence covering the period from 1961 to 1991 shows a very consistent continuation of this trend. Although there have been changes in the classification of some occupational groups, the overall trend from the census, and from government surveys such as the General Household Survey, shows that the socio-occupational skill structure of the economically active labour force is undergoing a continuous long-term process of upward shift. The professional, managerial and technical groups (male and female) are increasing in number and in proportionate importance, the intermediate and junior non-manual group is stable or declining and *all* the manual groups – skilled, semi-skilled and unskilled – are in continuous long-term decline, not just the latter two groups as Glass (1973) believed.

Whilst these data do not include those who have left the labour force or ceased to be economically active, they do include the unemployed and in this respect they provide the best available picture of what is happening. Two points should be noted however. First, there is clear evidence of systematic under-enumeration in the 1991 census, particularly among young males, and those from ethnic minorities in the inner areas of the major cities. This was particularly marked in London (Simpson and Dorling, 1994;

Hall and Hall, 1995), and it has been estimated that perhaps as many as 10 per cent of males in their twenties were not enumerated in 1991. As a high proportion of this group are likely to have been either unemployed or employed in less-skilled jobs, this may well have led to some undercounting of the size of the less-skilled workforce. Second, it is also argued that there is now a substantial workforce of illegal immigrants in London which has grown substantially in the 1990s. This group are likely to exist outside the boundaries of official data collection and tend to be paid 'cash in hand' for their work. These problems notwithstanding, it is argued that official statistics provide us with the best and most comprehensive picture of large-scale economic and social change. If we discount all official statistics on the grounds that they are flawed or partial, we are left with, at best, a very fragmentary picture derived from small-scale informal surveys, or at worse, pure surmise, speculation, anecdote and personal observation.

The transformation of London's industrial structure from manufacturing to financial, business services and the creative industries has been associated with an equally dramatic transformation of London's occupational and class structure. Put simply, the shift from an industrial to a post-industrial city has been paralleled by a shift from an industrial to a post-industrial occupational structure. This shift has, of course, taken place across Britain as a whole (Marshall and Rose, 1985), along with all developed western economies, but it has been marked in London and the South East, where the transition from an industrial economy to one based on business and financial services has been most rapid (Graham and Spence, 1995).

The changing socio-economic structure of London, 1961–91

Sassen (1991) argues that: 'New York, London and Tokyo show parallel employment and earnings trends' (p. 24), and suggests that, while London has witnessed an increase in professionals and managers, 'the number of clerical and blue-collar service jobs has also increased significantly in the City and in London at large' (p. 269). Although she gives data on the changes in the industrial structure of London, she provides no data to support her claim that London has experienced an 'occupational transformation' (p. 233). Indeed, there is no data on occupational change on London anywhere in her book. This section therefore examines socio-economic data to see what evidence there is, if any, for polarisation. As it will show, the empirical evidence for London, and for other major global cities, runs directly counter to the polarisation thesis.

The data used here are derived from published data for the 1961, 1971, 1981 and 1991 censuses on socio-economic classification of economically active usual residents, both employed and unemployed. There are seventeen

Table 3.1 The standard seventeen socio-economic groups

1	Employers and managers in large organisations
2	Employers and managers in small organisations
3	Professional workers, self-employed
4	Professional workers, employees
5.1	Ancillary workers and artists
5.2	Foremen and supervisors, non-manual
6	Junior non-manual workers
7	Personal service workers
8	Foremen and supervisors, manual
9	Skilled manual workers
10	Semi-skilled manual workers
11	Unskilled manual workers
12	Own account workers, other than professional
13	Farmers – employers and managers
14	Farmers – own account
15	Agricultural workers
16	Members of Armed Forces
17	Inadequately described occupations

Source: Office of Population, Censuses and Surveys.

socio-economic groups and these are shown in Table 3.1. From these groups, the standard sixfold classification is derived. The first group (1, 2, and 13) consists of employers and managers; the second (3 and 4) of professional workers, both employers and employees; the third group (5 and 6) consists of other non-manual workers (artists and allied workers, intermediate and junior non-manual workers, including clerical workers). The fourth group consists of skilled manual and supervisory workers and self-employed non-professional workers (8, 9, 12, 14). The fifth group comprises semi-skilled and personal service workers (7, 10, 15). Finally, there are unskilled manual workers (11). In addition, there are two residual categories – the Armed Forces (16) and 'occupations inadequately described' (17). All socio-economic data are based on a 10 per cent sample of enumerated population, and therefore suffer from sampling errors, though with the numbers and geographical areas involved these are so small as to be insignificant.

It should be stressed that a major change in the classification of occupations occurred in 1980 which affected the socio-economic group classification employed in the 1981 census. This had the effect of reducing the proportion of professionals, increasing the proportion of intermediate non-manual workers, reducing the proportion of junior non-manual workers and raising the proportion of semi-skilled and personal service workers. No adjustments have been made to the data to take account of these changes, which have affected groups at the margins. It should also be noted that a new category was added to the 1991 census, namely those on government training schemes. In this analysis they are counted alongside groups 16 and 17. These groups form a residual category of limited numerical importance and they can be disregarded here. Detailed analysis of the changes have been published elsewhere (Hamnett, 1994b, 1996) and we will focus here on the broad changes from 1961 to 1991.

A second, and related, problem is that, whilst in Britain the seventeen standard socio-economic groups have remained the same, the occupations assigned to them have not. In the 1970 and 1980 Classification of Occupations, the Registrar-General reassigned a number of occupations to different, and usually lower, SEGs for broadly the same reasons as those identified by Braverman and Kumar. This particularly affected the 1980 classification and hence the 1981 census. In an attempt to assist comparison over time, OPCS reclassified a national sub-sample of 1 per cent of economically active males in employment from the 1971 census, by the SEG on the basis of the 1980 Classification of Occupations. Comparison of the proportion of males in this sub-sample classified according to both the 1970 and the 1980 classifications reveals significant changes, but they are not sufficient to undermine the general findings presented here.

Looking first at the occupational structure of all economically active males in Greater London in 1981 and 1991, Table 3.2 shows there was an absolute decline of 12 per cent in the number of economically active males over the period: largely a product of the severe recession of the early 1980s which hit manufacturing employment particularly hard. In this context, there was an absolute increase in the number of professionals (13 per cent) and managers (11 per cent), and a decline in the size of all other groups. There was a 5 per cent decline in the number of intermediate and junior non-manual workers, a 19 per cent fall in skilled manual workers, a 24 per cent fall for semi-skilled and personal service workers, and a 32 per cent fall in number of

Table 3.2 Socio-economic change in Greater London, 1981–91: economically active males, employed and unemployed

Grouped SEGs	1981	1991	Absolute change	Change (%)	1981 (%)	1991 (%)	1981–91 (ppc)
3, 4	12,697	14,316	1,637	12.9	6.4	8.2	1.8
1, 2, 13	31,961	35,472	3,511	11	16.1	20.4	4.2
5, 6	43,528	41,361	−2,167	−5	22	23.7	1.7
8, 9, 12, 14	59,809	48,142	−11,667	−19.5	30.2	27.6	−2.6
7, 10, 15	28,377	21,635	−6,742	−23.8	14.3	12.4	−1.9
11,	11,402	7,739	−3,663	−32.1	5.7	4.4	−1.3
16, 17, gov.	10,225	5,630	−4,595	−10.2	5.1	3.2	−1.9
Total	197,981	174,295	−23,686	−12	100	100	0

Sources: Censuses of Population, 1981 and 1991, Economic Activity Tables (10% data).

Table 3.3 Socio-economic change in Greater London, 1981–91: economically active females, employed and unemployed

Grouped SEGs	1981	1991	Absolute change	Change (%)	1981 (%)	1991 (%)	1981–91 (ppc)
3, 4	2,376	4,576	2,200	92.59	1.7	3.3	1.6
1, 2, 13	10,560	18,598	8,038	76.12	7.6	13.5	5.9
5, 6	81,408	80,071	−1,337	−1.64	58.5	58.3	−0.2
8, 9, 12, 14	6,239	6,072	−167	−2.68	4.5	4.4	−0.1
7, 10, 15	24,991	17,995	−6,996	−27.99	18	13.1	−4.9
11,	7,100	6,771	−329	−4.63	5.1	5.1	−0.2
16, 17, gov.	6,429	3,303	−3,126	−48.62	4.6	2.4	−2.2
Total	139,103	137,386	−1,717	−1.23	100	100	0

Sources: Censuses of Population, 1981 and 1991, Economic Activity Tables (10% data).

unskilled workers. The proportion of managers rose by 4.2 percentage points (from 16.1 per cent to 20.3 per cent), the proportion of professionals rose by 1.2 points, and intermediate and junior non-manual workers by 1.7 points. All three manual groups declined in proportionate terms and the lower the skill group the greater the magnitude of the decline. There is no evidence whatsoever of polarisation and a strong trend towards what can be termed 'professionalisation'.

Changes in the socio-occupational structure of economically active females (Table 3.3) are broadly similar, although the structure of women's occupations is very different from that of men. Whereas the largest single male occupational group is that of skilled manual workers, by far the largest single female group is that of intermediate and junior non-manual workers which accounted for some 58 per cent of all economically active women in both 1981 and 1991.

The number of women in professional and managerial jobs grew very rapidly, by 93 per cent and 76 per cent respectively, though admittedly from a very low initial base. This indicates that women had begun to move into managerial and professional jobs, and there is evidence that this process has continued in the 1990s: 50 per cent of all trainee solicitors are now women. The proportion of women in all other occupational groups fell – by a remarkable 28 per cent in the case of semi-skilled and personal service workers, the very group which might have been expected to grow according to polarisation theorists. The trends for Inner London are shown in Figure 3.1.

There are several possible counters to this evidence. The first is that the reality of social polarisation is obscured by the aggregation of different socio-economic groups together. The second is that polarisation may not have manifested itself in the 1980s but did so earlier. A third is that it may exist at a more local level though not at the aggregate city scale. This argument

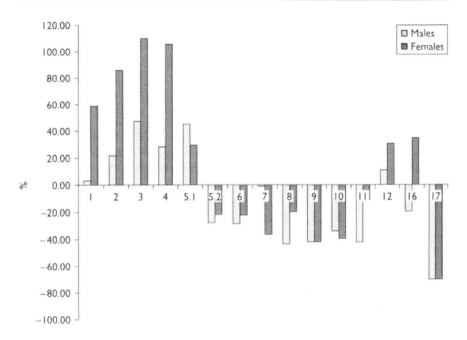

Figure 3.1 Percentage SEG change, Inner London, males and females (excl. agriculture), 1981–91.
Sources: Censuses of Population, 1981 and 1991.

seems to contradict the polarisation thesis which is essentially city-wide in application. Finally, Bruegel (1996) argues that polarisation is now taking a gender-specific form whereby certain sectors are being simultaneously feminised and deskilled and downgraded in status.

Taking the counter arguments one by one, none of them are supported by evidence. If we look at changes in individual socio-economic groups for Greater London 1981–91, all the socio-economic groups in the professional and managerial groups have grown in absolute terms as has SEG 5.1 (artists and ancillary workers) (Figure 3.1). All the other SEGs have decreased in importance with the exception of socio-economic group 12 (the self-employed), which reflects the growing trend towards self-employment in recent decades. In addition, the percentage growth of the professional and managerial groups has been more marked for women that for men in every case. The feminisation of these groups is occurring, but it is unclear why this should involve deskilling.

The second potential counter argument is that polarisation may have occurred prior to the 1980s. Again, there is no evidence for this. On the contrary, Figures 3.2 and 3.3 show that growth of the managerial and professional SEGs has been taking place since the 1960s. The proportion of

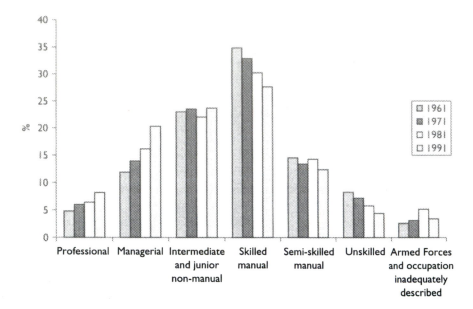

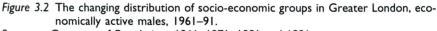

Figure 3.2 The changing distribution of socio-economic groups in Greater London, economically active males, 1961–91.
Sources: Censuses of Population, 1961, 1971, 1981 and 1991.

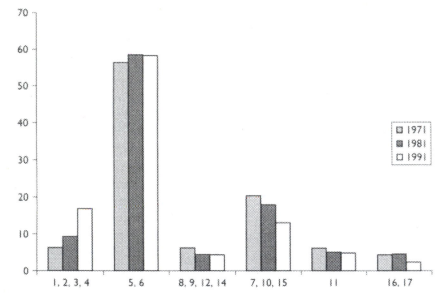

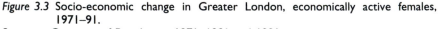

Figure 3.3 Socio-economic change in Greater London, economically active females, 1971–91.
Sources: Censuses of Population, 1971, 1981 and 1991.

economically active professional males rose in each intercensal decade, from 4.8 per cent in 1961 to 8.2 per cent in 1991, while the proportion of managers rose in each intercensal decade from 12 per cent in 1961 to 20.3 per cent in 1991. Professionals and managers together accounted for 28 per cent of economically active males in 1991 compared to 16.7 per cent in 1961. The same pattern is true for females, though the increase is more marked in the 1980s. On the basis of census data on the economically active, there is also no evidence for the view that occupational polarisation has increased in London. On the contrary, the evidence points to a process of professionalisation in that the professional and managerial groups have steadily become more important within the occupational structure whilst intermediate and junior non-manual workers have remained stable and all the manual groups have proportionately decreased. The only counter-evidence is the growth of unemployment and economic inactivity since 1981 (Burgers, 1996), which may be seen to indicate a form of polarisation between those in employment and those who are long-term unemployed, or have left or are excluded from the labour force. This is a very different phenomenon, however, from that suggested by Sassen which focuses on polarisation of the economically active labour force. This is more akin to the process envisaged by Esping-Andersen of a division between labour market insiders and those excluded from the labour market, which is arguably more characteristic of Europe than of the USA.

Buck reaches similar conclusions in his comparison of Labour Force Survey data for London residents from 1979 to 1996. Following Esping-Andersen's division of occupations into 'Fordist' and post-industrial sectors, he concludes that:

> The 'fordist' occupations display declining shares, especially so amongst manual workers. Within this hierarchy there is a marked growth of the top, and a decline at the bottom, though in total this segment is declining. Within the 'post-industrial' hierarchy, growth is again concentrated at the top, and is particularly strong amongst semi-professional workers. Most notably, there is no growth in the share of 'unskilled service workers' – the service proletariat . . . In summary neither hierarchy shows polarisation, in the sense of growth at the top and bottom at the expense of the middle. These findings tend to support those of Hamnett (1994) . . . and confirm that there is no occupational polarisation . . . but rather something closer to professionalisation.
>
> (Buck, 1997, pp. 8–9)

Buck also finds little evidence to support the idea that the London labour market has become disproportionately 'flexible', in the sense of having a large and growing share of temporary or part-time jobs. Although the proportion of this type of job has grown in both London and the UK from

1986 to 1996, London has slightly smaller proportion of the labour force in this type of job than the UK as a whole in both years. Buck thus concludes that the evidence 'does not support the global cities model of occupational polarisation, or indeed a polarisation by job type into more and less secure jobs' (p. 12).

Summary and conclusions

London is not alone in experiencing these changes in its occupational class structure. Similar changes have occurred in Amsterdam (Hamnett, 1994a), in Paris (Preteceille, 1995; Rhein, 1998); in New York (Brint, 1991; Mollenkopf and Castells, 1991), San Francisco (Simmie, 1983), Singapore (Baum, 1999), Toronto, Vancouver and other Canadian cities (Ley, 1996). The dominant experience of global cities seems to be one of an upwards shift in the occupational class structure of the economically active labour force rather than polarisation. Writing on New York, Brint noted that:

> where salaried professionals and managers . . . accounted for something under 5 percent of the working population throughout the nineteenth and early twentieth century, they now represent nearly 30 percent of the labour force [of New York]. At the higher end of the professional and managerial spectrum, affluent professional people are now so numerous, visible and influential that some observers have characterised them as the 'new dominant class' of the postindustrial city.
>
> (Brint, 1991, p. 155)

Brint's observations are paralleled by the evidence in London. Data from the census of 1891 for London show a total of 6 per cent of males and under 3 per cent of females in professional (including teaching) and administrative occupations (Stedman Jones, 1971). In social-class terms, the 1891 census showed 6.5 per cent of males and 5.3 per cent of females in class 1. This compares to a figure of almost 22 per cent of males in unskilled occupations. It is clear that there has been a dramatic change in social-class composition in both cities. Whereas the social commentators of the nineteenth century pointed to the flight of the middle classes from parts of central and Inner London, the dominant process today is not out-migration of the middle classes, but gentrification. Interestingly, Sassen comments of London that:

> The most central areas of London have undergone a transformation that broadly parallels Manhattan's . . . We see a parallel increase in the stratum of what Brint (1991) has described as upper professionals, a group largely employed in corporate services, including finance. The sharp growth in the concentration of mostly young, new high-income

professionals and managers employed in central London represents a significant change from a decade ago.

(Sassen, 1991, p. 265)

The argument made here is that this occupational transformation has not been solely confined to the most central areas of London. On the contrary, it is characteristic of both Inner and Outer London, and of Greater London as a whole. As a consequence, what remains of the traditional manual working class is being gradually replaced in much of Inner London through a process of long-term gentrification. In the process, the less skilled and the economically inactive have become increasingly concentrated in an increasingly residualised public housing sector. These processes of residential and housing market change are discussed in Chapters 6 and 7 respectively, and some of the social consequences in terms of the geographical distribution of deprivation and social exclusion are discussed in Chapter 8.

Chapter 4

Widening inequality

Earnings and incomes

> like the golden cities of ancient empires, they draw unto themselves the wealth of the world that is ruled by them. They become the major points for the accumulation of capital and 'all that money can buy'. They are luxurious, splendid cities whose very splendour obscures the poverty on which their wealth is based. The juxtaposition is not merely spatial, it is a functional relation: rich and poor define each other.
>
> (Friedmann and Wolff, 1982, p. 319)

> It is clear that the number of well-off people in the City increased quite rapidly in the 1980s as a result of generally improved levels of income, combined with a gradual increase in the number of professional and managerial workers who were the chief beneficiaries of these increases in income, both relatively and absolutely.
>
> (Thrift and Leyshon, 1992, p. 291)

Introduction

It is clear from previous chapters that London has undergone a rapid transformation in recent decades from an industrial city, with a third of its workforce in manufacturing, and a large number of skilled and semi-skilled manual workers, to a post-industrial city dominated by financial and business services employment with a predominantly non-manual workforce and a growing proportion of professional and managerial workers. This chapter examines the implications of this transition for the distribution of both *individual earnings* and *household incomes*. It is important to differentiate the two as earnings data relate to earnings from employment, and are usually available only for individual earners and, by definition, exclude the unemployed who have no earnings from formal work. Income comes from a variety of sources, including earnings, but it also includes unemployment pay and other transfer payments such as pensions, and supplementary benefits, interest and dividends. Most income data are for households, which is important, as questions of income inequality are particularly salient at the household level (Williams and Windebank, 1995).

The key questions addressed in this chapter are to what extent has London become a more unequal or polarised city in terms of earnings and incomes over the last twenty years; to what extent has it diverged from the rest of Britain in terms of inequality; and how far are the trends linked to changes in the industrial and occupational structure of London, or to what extent are they simply reflections of national trends? At one level the evidence is clear: the proportion of people driving around in luxury cars, the boom in expensive bars and restaurants, and the soaring house prices which have raced ahead of the rest of the country in the second half of the 1990s, all point to the fact that there is a lot of money in London and a high level of conspicuous consumption to match. A key characteristic of the high incomes and wealth in global cities is their role as centres of conspicuous consumption. Today, as in the past, whole industries have sprung up to cater for the consumption of the rich. The luxury car showrooms, country house estate agents (Thrift and Leyshon, 1994), expensive restaurants, fine art dealers, couturiers and interior designers could not exist without them and, like it or not, they play an important part of the economy of great cities.

But perhaps the emblematic event which defined conspicuous consumption in London around the turn of the new century was the bill run up by six investment bankers for a meal at the restaurant Petrus. The bill for the six was £44,000, made up of three bottles of Petrus vintage claret 1945, 1946 and 1947, at an average cost of £11,000 per bottle, followed by a bottle of 1900 Château d'Yquem costing £9,200, and a bottle of 1982 Montrachet at £1,200. Adding on a few pounds for bottles of water and two beers, the restaurateur threw in the £300 cost of the meal free (*Sunday Times*, 15/07/01; *Guardian* 10/07/01).

What resonated so strongly in Britain with this story, and why it received so much media publicity, was the dramatic light it threw on the levels of disposable income of highly paid City workers. Buying a house, luxury car or holiday with the proceeds of a major deal would be understandable. But £11,000, even for a vintage Château Petrus, beggars belief, as does the realisation that there are people around who can afford to pay this much for a bottle of wine. But pay they can, and this is at the heart of the issue. It was leaked that the six bankers (all males between 25 and 35) were all employees of Barclays Capital, the investment banking division of Barclays Bank, out celebrating completion of a big deal. The Bank denied they had been there on official business and said it was a private night out and they were paying themselves, but the event is a telling indication of the level of disposable income in the City.

But London's wealthy elite are only one side of the coin. At the other end of the spectrum, the deep recessions of the early 1980s and early 1990s hit London hard and led to a major rise in unemployment from 200,000 in 1985 to 464,000 in 1988. The rise in unemployment disproportionately hit the less skilled. Unemployment among males over 50 and under 20, and in some ethnic minorities, rose particularly fast. Most of the unemployed are on state benefits and low incomes, and some parts of London are among the most deprived in Britain with high proportions of social security recipients (London Research Centre, 1996; Harrison, 1983; Townsend, 1987).

Income inequality and the national and international division of labour

There is no doubt that London, like other global cities such as New York (Mollenkopf and Castells, 1991) and Paris (Preteceille 2001), is a very unequal city – all have a large concentration of both high earners and high-income households who command a very disproportionate share of their cities' wealth. This is a direct consequence of their role as centres of finance capital and corporate control, and the residence of many of the national and international wealthy elite, but it is not a new phenomenon and nor is the concentration of poverty and low incomes in London as Stedman Jones (1971) and Green (1986) have shown.

London has long possessed the greatest concentration of high incomes in Britain. This can be traced back to the location of the court in London, which led to a concentration of the aristocracy, high incomes and conspicuous consumption. More recently this reflects the economic and commercial dominance of London during the nineteenth century. Rubenstein's (1977) analysis of the geographical distribution of wealth in nineteenth-century Britain demonstrates that London was the major centre of wealth in Britain and that the majority of large fortunes were earned by merchants, bankers, ship owners, and stock or insurance brokers living and working in the City rather than by manufacturers and industrialists. As Rubenstein put it:

> not merely were large fortunes earned more readily in London, but the highest peaks of wealth were reached by Londoners especially by City men. In contrast, the fortunes earned in industrial Britain were fewer in number and relatively smaller in size.
>
> (Rubenstein, 1977, p. 108)

Rubenstein showed that the concentration of wealth in London was not limited to the very wealthy. The income of the Victorian middle class also came disproportionately from London. Not only did London possess a larger taxable business income than all the chief provincial towns combined, but income per capita was larger. Similar work by Green (1995) has shown the significance of the middle classes in nineteenth-century London using data on inheritance. Rubenstein concludes that there were important differences between London and the provincial-based middle class in terms of the role they played in the national spatial division of labour. London functioned as the control centre of the economy. As he put it:

> the distinction between commercial wealth, based largely in London, and manufacturing wealth, based in the provinces, is not the artificial product of geography but a basic dichotomy within the nineteenth century middle class as a whole. The chief economic distinction between the London and provincial based middle classes was that one was largely capital- and the other largely labour-intensive.
>
> (Rubenstein, 1977, p. 112)

Although Rubenstein's analysis is not without critics (Mitchie, 1992), the implication is that the dominance of finance capital in London is not a new phenomenon but is of long standing, although its quantitative importance has increased dramatically in the last fifteen years since Big Bang in 1986.

Growing inequality or polarisation?

It seems superficially clear that the gap between rich and poor has widened – but what do we mean by this, and what is the evidence for it? The conventional wisdom is that London, like other global cities, has become more polarised in terms of earnings and incomes; but there is an important distinction between polarisation and inequality. As Pahl (1988) recognises: 'there is an important distinction between images based on numbers of households or numbers of people and images based on the amount of wealth held by people at different levels' (p. 260). Thus *inequality* refers to the extent of dispersion between different levels of income/earnings, whereas *polarisation* relates to changes in the absolute and relative size of the groups in different income/earnings groups over time. These differences are often confused and greater income inequality is often taken as an indication of polarisation. But, although greater inequality and greater polarisation may be linked in practice they are analytically quite distinct, as Kloosterman (1996) and Esteban and Ray (1994) have shown. To put it simply, the distinction is whether the number of earners at both the top and bottom ends of the distribution is growing at the expense of the middle (polarisation) or whether the gap between the top and bottom of the distribution is growing (greater inequality). The key question is whether there are more rich and more poor (and less in the middle), or whether the rich are getting richer and the poor relatively poorer in relation to the rich. It is possible that both are happening, or one, or neither, but we need to be clear what we believe is occurring and what we are trying to measure.

One of the most influential advocates of polarisation is Sassen (1991) who argues, as outlined in Chapter 3, that global cities are characterised by growth at both the top and the bottom end of the occupational skill spectrum and an absolute decline in the size of the middle-skill group as a reflection of changes in the productive base of the city away from manufacturing industry (which is dominated by skilled and semi-skilled manual workers) and towards both business and personal services. It is argued that there are more professional, managerial and technical workers and more low-skilled service workers. She argues that the high-skill professional and business services employees are generally highly paid, and that the low-skilled personal service sector workers are very poorly paid. Consequently, there is a *polarisation* of earnings with a large and growing group of high earners and a growing group of low earners who service the transnational elite. In other words, the emphasis is on the size of various skill and earnings groups rather than simply the level of earnings inequality between different groups. Sassen (1991) claims that there has been 'a growth of a high income stratum and a low income stratum of workers' (p. 13) and she argues, following Noyelle and Stanbach (1985), that the growth of financial and business services jobs is linked with concentrations of incomes at both the top and bottom, whereas the manufacturing jobs they replace tended to be middle income. As she puts it:

> Major growth industries show a greater incidence of jobs at the high- and low-paying ends of the scale than do the older industries now in decline. Almost half the jobs in the producer services are lower income jobs, and half are in the two highest earnings classes. In contrast, a large share of manufacturing workers were in the middle earnings jobs during the post-war period of high growth in these industries in the US and the UK.
>
> (Sassen, 1991, p. 9)

To highlight the differences between notions of earnings polarisation and inequality may seem a trivial difference to some readers: mere definitional and academic nit-picking of little or no consequence. Nothing could be further from the truth. To say that there are both more rich and more poor (polarisation) is a very different thing from saying that the rich are getting richer and the poor are getting poorer (or richer but at a slower rate). We could, for example, find that London is becoming a city dominated by a large and growing high-income group, as Glass (1973) suggested. Conversely, we could find it is becoming dominated by lower-income groups, which was the fear in the 1970s (GLC, 1969; Eversley, 1972). Or, as the polarisation thesis claims, we could see simultaneous growth of both groups with the middle-income groups being squeezed out. Whatever interpretation is correct, the implications for employment, education, housing affordability and social cohesion are profound. The prospects raised by a city dominated by the better off, or by the poor or by both rich and poor, are very different.

This chapter will be attempting to argue four things using official data sources. First, that the incomes of the high-income groups have risen much faster than those of the poor, and that inequality has consequently increased dramatically in London since 1979. Second, that the relative size of the rich group has also increased substantially, but this is less true of the poor. In other words there has been a growth of inequality but not of polarisation. The size of the gap between rich and poor has grown, but there are not more low earners than there were thirty years ago if we look at inflation-adjusted earnings. The key to explaining this seeming paradox is that real inflation-adjusted earnings have risen for all groups, but they have risen much faster for higher-earning groups than for low. The third point is that the greatest increases in earnings have been amongst those working in the City of London, which points to the key role of financial and business services in increasing inequality. The fourth point is that the inequality in household incomes has risen substantially, and that both the incomes and the share of total household income of the lower groups have declined substantially while those of the richer households have increased. This is partly a result of changes in household composition, with more smaller households, and partly a result of a growing division between rich and poor households: particularly between two-earner households and no-earner households.

Measuring inequality and polarisation

Inequality is usually measured by calculating ratios to see how various parts of the distribution have changed relative to the median or the mean and how their shares have changed. For example, the inter-quartile ratio compares the earnings or incomes of the top quartile (25 per cent) to the bottom quartile (25 per cent) of the distribution and can be used to see if the ratio has changed over time. But, as polarisation relates to the size of various income groups, changes in such ratios will not capture the relevant trends. It is necessary to measure changes in the size of the top, bottom and middle of the earning/income distributions to see whether they have got bigger or smaller over time after adjusting for inflation.

There are two different methods used to determine whether polarisation is occurring. The first method, used by Kloosterman (1996), divides the earnings distribution into a low (first 25 per cent), the middle (50 per cent), and a high segment (top 25 per cent) for a selected year. The two cut-off points, expressed in money terms, are then transferred (correcting for the rate of inflation) to another later year. Subsequently the shift between the three segments can be calculated in terms of the numbers of wage earners. A rise in the proportion of earners in the top and bottom groups and a fall in the middle group over time is construed as evidence of polarisation. The second method, used by Harrison and Bluestone (1988), compares the changing proportions of three groups in relation to the median – those with earnings more than twice the median, those with under half the median, and the remainder. This has the advantage of taking the changing median into account, but it is also partly a measure of inequality.

Data sources: the New Earnings Survey and the General Household Survey

The data used in this chapter are drawn from both the New Earnings Survey (NES) and from the General Household Survey (GHS). In most of the analyses figures for 1979 and 1995 are compared. One of the great strengths of the NES is the size of the sample – nearly 1 per cent of all employees nationwide for a week in April. Over 154,000 individuals were included in 1995. Such a large sample means that sampling errors do not become significant until the data are disaggregated to a small spatial scale, and it is possible to analyse data for Westminster and the City of London as well as for Greater London, the Rest of the South East and Great Britain as a whole. The second strength of the survey is that it is selected from Inland Revenue PAYE records, and it is very comprehensive for employees who pay income tax. The Achilles' heel of the survey is that it excludes income from self-employment and does not collect any information on individuals with incomes below the tax threshold. This is particularly problematic for part-time workers who are often very low paid. The 1995 survey reported that:

> coverage of full-time adult employees is virtually complete but the coverage of part-time employees is not comprehensive. Many of those with earnings below the income tax threshold – in this survey equivalent to full-time earnings of £68 a week or £294 per month – are not covered, which excludes mainly women with part-time jobs and a small proportion of young people.
>
> (NES Survey, 1995, p. 1)

It has been estimated that about one-fifth of part-time employees, most of whom are women, are not covered by the survey for this reason, and this should be borne in mind when considering the results. But, notwithstanding these problems, the data on part-time workers do enable both absolute earnings and relative rates of change to be compared to those of 'workers in full-time jobs on adult rates of pay whose pay was not affected by absence'. This group totalled 69,057 males and 39,737 females in Britain in the 1995 survey. The equivalent figures for London were 10,188 and 6,595, and for the City of London they were 1,098 and 654 respectively.

The General Household Survey or GHS is also an annual survey, but of households rather than individual earnings, and it is a survey based on place of residence rather than workplace. The sample size for the GHS is much smaller than for the NES, and the number of households in Britain for which we have income data was 9,722 in 1979 and 8,814 in 1995. In Greater London the figures were 1,209 and 1,046 respectively. These are still reasonably large samples, but the GHS is subject to a larger degree of sampling error than the NES. On the other hand, GHS data include all earners, both full- and part-time, and those below the tax threshold, and it is a more inclusive source for earners than the New Earnings Survey. Critics argue that neither source will pick up those working in the informal economy outside the tax system (Cox and Watt, 2002). This is true, but as we have no data on these groups we can only find out what is happening at the bottom end by informal surveys or qualitative interviews with specific groups, and there is little or no systematic evidence.

The growing earnings gap between London and the rest of Britain

The distinction made by Rubenstein about middle-class wealth in nineteenth-century London being largely based on commerce and finance rather than industry remains true today. As Massey (1988) has argued, it is possible to identify a spatial division of labour in Britain in which control, command and financial functions are concentrated in Greater London, research and development activity in the wider South East, and routine manufacturing in the more peripheral regions of the country. If anything, the dominance of commercial and financial-based wealth in London has increased in the last twenty years and London has seen rapid growth in the number of high earners. The New Earnings Survey shows that both mean and median earnings in Greater London were higher than those in ROSE or GB in 1979, and the gap widened for almost all categories of employees between 1979 and 1995.

Table 4.1 shows that mean gross weekly pay for full-time male workers in Great Britain rose by 35 per cent in real terms from 1979 to 1995, compared to 47 per cent in the South East and 57 per cent in Greater London. For full-time females, earnings rose by 56 per cent in GB, 65 per cent in the South East and by 73 per cent in Greater London. Earnings of female part-time workers rose by 45 per cent in GB, 56 per cent in the South East and 61 per cent in Greater London. Increases in median earnings are lower but show the same general pattern. Thus, whereas male mean weekly pay in Greater London was 14 per cent greater than for GB overall in 1979, in 1995 it was 33 per cent higher. The equivalent figures for full-time females were 16 per cent and 29 per cent respectively. The differences in median earnings were lower but show the same pattern. It is clear that earnings in Greater London increased at a much faster rate than in Britain as a whole and the rest of the South East. This reflects London's role as a global city and its industrial and occupational composition (see also Buck, 1997, table 7).

Table 4.1 Percentage increase in gross weekly pay, 1979–95 (at April 1995 prices) for Great Britain, the South East and Greater London

	Male full-time workers			Female full-time workers			Female part-time workers		
	GB	SE	GL	GB	SE	GL	GB	SE	GL
Top decile	48.6	62.4	73.1	69.5	76.5	83.4	82.3	97.3	97.2
Top quartile	37.3	48.3	55.9	66.4	72.9	77.9	45.9	55.8	65.5
Mean	34.6	46.9	57.2	56.1	65.5	73.5	44.8	55.7	60.9
Median	25.4	34.4	41.1	48	58.7	46.5	28.9	35.3	44.1
Bottom quartile	14.6	22.2	27	36.6	46.5	54.7	16.2	22.9	27.8
Bottom decile	7.1	11.7	17.8	26.6	36.1	45.4	10.8	14.3	16.7

Sources: New Earnings Survey, 1979 and 1995.

The increases in earnings have also been accompanied by an increase in inequality. Table 4.1 shows that earnings in both the top decile and the top quartile rose far more rapidly than in the bottom decile and quartile. In Great Britain as a whole, earnings for the top quartile of male full-time workers rose by 37 per cent compared to 15 per cent for the bottom quartile, whereas earnings in the top and bottom deciles rose by 49 per cent and 7 per cent respectively: a difference of 7:1. The same pattern holds for both full-time and part-time females and indicates a considerable increase in earnings inequality over the period. In the South East and in Greater London the discrepancy between increases in the earnings of full-time male workers in the top and bottom deciles and the top and bottom quartiles was less marked. Although earnings of the top decile in Greater London rose by a remarkable 73 per cent over the period, the earnings of the bottom decile rose by 18 per cent: a ratio of 4:1. For female full-time employees, earnings in the top decile increased by a remarkable 83 per cent, but even in the bottom decile it was still 45 per cent – much higher than the lowest decile of male workers.

It is important to stress here that while there was a sharp increase in inequality in Britain as a whole the earnings of the bottom decile still rose in real terms. This is very different from the situation in the United States where earnings of the lower groups have fallen in real terms in recent years. This reflects the fact that there is a less-low-paid service work in Britain than in the USA and the labour market is less 'flexible'. Put another way, it is more difficult in Britain than in the USA to get away with paying very low wages.

The increase in both mean and median earnings was greater for women than for men, and was higher in Greater London than in either the South

East as a whole or in Great Britain. This applies to both female full-time and part-time workers. London was the motor of high earnings growth for both men and women in Britain over this period. The discrepancy is greatest at the bottom of the earnings distribution where female full-time earnings rose in GB, the South East and in Greater London by three times the rate of earnings increase for males. At higher levels the gap gets progressively smaller, though it is noticeable in the top decile. It should be pointed out, however, that men still earned more than women at every level in the earnings distribution in 1995, notwithstanding the reduction in the size of the gap. The male–female earnings gap has been narrowed, but it has by no means disappeared. This argument that the growth of inequality is driven by earnings growth at the top is strongly supported by Gordon, who notes that:

> Earnings inequalities in the London labour market are greater than in the country as a whole, although this is almost entirely due to the top tail of the earnings distribution among male non-manual workers (where the top decile was 75 per cent above the average, and 4.3 times the lowest decile, compared with 3.7 times nationally). In other words, it reflects the presence of a disproportionately large number of very highly rewarded professional and managerial workers, rather than any larger number of poorly paid jobs.
>
> (Gordon, 2000, pp. 19–20)

The impact of these differential changes in earnings can be summarised using inter-decile and inter-quartile ratios. Table 4.2 shows that earnings inequality was more marked in London in 1979 than in the South East or GB as a whole: the inter-decile ratio for male full-time workers in 1979 was 2.4 in GB, 2.5 in the South East and 2.6 in London. By 1995 these ratios

Table 4.2 Changes in inter-quartile (IQ) and inter-decile (ID) earnings ratios, 1979–95

	Male full-time workers				Female full-time workers				Female part-time	
	IQ ratio		ID ratio		IQ ratio		ID ratio		ID ratio	
	1979	1995	1979	1995	1979	1995	1979	1995	1979	1995
Great Britain	1.56	1.87	2.38	3.3	1.52	1.85	2.28	3.05	1.79	2.94
South East	1.6	1.95	2.5	3.64	1.53	1.81	2.29	2.98	1.9	3.28
Greater London	1.62	2.01	2.6	3.82	1.48	1.7	2.26	2.85	1.98	3.34
GL/GB	1.04	1.07	1.09	1.16	0.97	0.92	0.99	0.93	1.11	1.14

Sources: Calculated from NES data, 1979 and 1995.

had risen to 3.3, 3.64 and 3.82 respectively. This shows that earnings inequality rose across the country as a whole, but that it was particularly marked in London. For full-time female employees the picture was rather different. Earnings inequality increased across Britain as a whole but it was slightly lower in London than in Britain. This may reflect a greater demand for female labour in London. A similar picture is seen for manual and non-manual earnings, although non-manual workers saw much greater rises than manual workers in virtually every category.

Earnings differentials within London: the City and Westminster versus the rest

Sassen (1991) argues that the key determinant of global city economic status is the key role played by financial and business services rather than corporate headquarters. She also argues, along with Friedmann and Wolff (1982) and Mollenkopf and Castells (1991), that the financial and advanced business service sectors are among the major motors of increases in earnings inequality in global cities. This point is also forcibly made by Thrift and Leyshon (1992) who point to the massive earnings increases in the City of London in the run up to the 'Big Bang' in the late 1980s. Given this, we would expect to find that earnings are higher in the City of London and that they have risen faster than elsewhere. This is borne out by the analysis of NES data for both the City of London and for the City of Westminster which it was possible to compare to that for London as a whole. Table 4.3 shows that mean earnings rose by substantially more for both males (100 per cent) and females (102 per cent) in the City of London than they did in London as a whole (57 per cent and 73 per cent respectively). This was also true for the top and bottom deciles of earnings. Earnings for the top decile of male full-time workers rose by 136 per cent in the City compared to 73 per cent in Greater London, and female full-time earnings rose by 117 per cent compared with 83 per cent. The figures for Westminster are

Table 4.3 Percentage increases in gross weekly pay, 1979–95 (at April 1995 prices) for Greater London, Westminster and the City of London

	Male full-time workers			Female full-time workers		
	GL	West.	City	GL	West.	City
Top decile	73.1	68.1	135.8	83.4	93.7	116.6
Mean	57.2	65.6	99.9	73.5	73.6	102.3
Bottom decile	17.8	23.7	43.1	45.4	48.5	74

Sources: Calculated from NES data, 1979 and 1995.

Table 4.4 Inter-decile earnings ratios for full-time workers, 1979–95

	Male full-time workers			Female full-time workers		
	GL	West.	City	GL	West.	City
1979	2.6	3.17	2.98	2.26	2.13	2.11
1995	3.82	4.3	4.9	2.85	2.78	2.62
Change	1.22	1.13	1.92	0.59	0.65	0.51

Sources: Calculated from NES data, 1979 and 1995.

higher than the Greater London figures but much lower than those for the City. It is clear that the City has led the upwards growth in earnings in London. To this extent, it is also a major driver in the growth of earnings inequality.

Not surprisingly, Table 4.4 shows that the inter-decile ratios for West-minster and the City were higher in 1979 than for London as a whole and that the increases over the period 1979–95 were also greater. The increases in the ID ratio were greatest in the City primarily as a result of the massive increases in the top decile of earnings. The male ID ratio in the City rose from 2.98 in 1979 to 4.9 in 1995.

The growth of earnings inequality within the service sector

A central element of the polarisation thesis advanced by Sassen is the degree of earnings polarisation within the service sector compared with that in manufacturing industry. It is argued that the service sector contains a higher *proportion* of both high and low earners than manufacturing, with a high *proportion* of workers in the low-paid catering, cleaning, and personal services sector, and a high proportion in the high-paid financial and business services sector. This proposition is difficult to examine empirically as it requires data on earnings distribution by industrial sector. Hamnett and Cross (1998a) examined changes in real earnings in Greater London by two major service industry divisions and found that there is a major gap between manual and non-manual weekly pay. Male non-manual pay in banking, finance, insurance and business services in London rose by far more than pay in the 'other services' groups, both of which rose by far more than pay for the manual groups.

Second, we examined changes in earnings in two occupational orders. The first group consisted of occupational group II: 'professional and re-lated, supporting management and administration'. The second occupational

Table 4.5 Absolute and percentage change in average gross weekly pay of occupational groups representative of full-time managers and personal service workers, 1979–95

	Male managers		Male cleaners		Female managers		Female cleaners	
	GL	GB	GL	GB	GL	GB	GL	GB
Average gross weekly pay (£ at April 1995 prices)								
1979	404	354	221	204	293	260	167	140
1990	655	528	251	218	473	397	205	166
1991	645	528	283	249	441	361	205	167
1995	708	567	288	252	480	399	203	171
Percentage change								
1979–90	62.4	49.2	13.4	7.3	61.3	53	22.6	18.2
1991–95	9.8	7.3	1.9	1.2	8.7	10.2	−0.6	1.9
1979–95	75.3	60.1	30.2	24	63.6	53.7	21.7	21.0

Source: Hamnett and Cross (1998a).

group consisted of 'catering, cleaning, hairdressing and other personal services'. The groups are consistently defined from 1979 to 1990, but from 1991 onwards managers are represented by group 1a 'corporate managers and adminstrators' and the second by 'other elementary occupations'. To take account of the definitional changes we split the analysis of earnings change into two time periods: 1979–90 and 1991–5.

Not surprisingly, we found evidence of sharp differences in earnings growth between the two groups. In Britain as a whole, the earnings of male managers rose by 60 per cent in real terms between 1979 and 1995, whereas the earnings of personal services rose by just 24 per cent (Table 4.5). In London, earnings and earnings increases were both greater, rising by 75 per cent for managers and 30 per cent for personal services. Increases for females were lower in all cases, although the same pattern is seen. Female managers had increases of 64 per cent in London and 54 per cent in Britain compared to increases of 22 and 21 per cent respectively for personal services. Although the labels are crude and schematic, the differences between the sectors are very clear. The wages of managers and administrators rose by far more than those of personal service workers.

Earnings polarisation or inequality?

There is strong evidence for the growth of earnings inequality in London over the period 1979–95, and it is very likely that this has continued and may have intensified since 1995 as 'bonus culture' has spread to the City from the USA and as earnings have grown sharply in response to the 'internationalisation' of salaries. This means paying City workers salaries akin to what their equivalents would earn in New York. As is argued below, the growth of inequality has been driven both by the growth of earnings (both salaries and bonuses) at the top end of the scale and by the ability of employers to offer lower wages at the bottom end of the scale where competition for jobs is strong and the workers' bargaining position is weak. This should be differentiated from the situation in the USA where earnings at the bottom have fallen in real terms.

The question addressed in this section, however, is the related one of whether there has been a polarisation of earnings and incomes in the specific sense of whether the size of the top and bottom of the earnings distribution has grown at the expense of the middle. In other words, the question is 'are there more rich and more poor and fewer in the middle groups?'. This is a difficult issue to try to measure. The method used splits the distribution of earnings in quartiles at a given point in time (1979), then adjusts the cutting points for inflation and applies adjusted 'cut' points to the distribution at a later point in time. Doing this it is possible to see what proportion of earners fall into the various inflation-adjusted bands in 1995. Two different sources of data are used: the GHS and the NES. First we look at the GHS.

The polarisation of earnings, 1979–95: the GHS data

The GHS data for London reveal a marked upwards shift in the earnings distribution from 1979 to 1993. Adjusting 1979 earnings thresholds to 1995 prices, the proportion of earners in what was the bottom quartile band in 1979 fell by 4 points in London; the proportion of earners in the middle fell by 21 points and the proportion in the top band rose by 25 points. The shift was far more marked in London than in Britain as a whole, where there was a redistribution of 13.9 percentage points from the middle to the top group. This reflects the fact that earnings rose much more rapidly in London than in Britain as a whole. This shift is even more marked at the decile level. Table 4.6 shows that, with the important exception of the bottom decile of earnings which grew by 1.6 points, the proportions in all other deciles to the eighth fell substantially. There was then a small increase in the proportion in the ninth decile and a massive increase (+22.4 points) in the 1979 top decile earnings band adjusted to 1995 prices. What the results show is a massive upwards shift in real earnings, particularly in London, away from all deciles except the bottom one, to the highest decile band. This shows limited polarisation in that there are more rich and more poor, but it is a very asymmetric form of polarisation, with an increase at the top end fifteen times that at the bottom.

Table 4.6 Changing proportions of earners by deciles, 1979 and 1993
(earnings adjusted to April 1993 prices using RPI)

Deciles	Gross weekly individual earnings in 1979 £s	Size of group (%)		Percentage point change 1979–93	
		1979	1993		
London: decile cutting points					
1	61.4	10	11.6	1.6	1.6
2	103.51	10	5.5	−4.5	
3	141.85	10	6.6	−3.4	
4	170.21	10	5.2	−4.8	
5	195.16	10	5	−5	−24
6	219.81	10	4.5	−5.5	
7	250.35	10	7.4	−2.6	
8	286.77	10	6.9	−3.1	
9	355.25	10	14.9	4.9	
10		10	32.4	22.4	22.4
Britain: decile cutting points					
1	44.39	10	11.1	1.1	1.1
2	79.55	10	9.8	−0.2	
3	115.30	10	7.8	−2.2	
4	145.95	10	7.1	−2.9	
5	172.91	10	6.8	−3.2	−16.1
6	200.33	10	6.7	−3.3	
7	230.50	10	7.7	−2.3	
8	267.31	10	8.4	−1.6	
9	325.74	10	9.6	−0.4	
10		10	25.0	15.0	15.0

Sources: General Household Survey, 1979 and 1993.

It could be argued that the RPI is not the most appropriate measure (although it does standardise in terms of purchasing power) and that it is better to adjust categories by using median income change for Great Britain. This is done in Table 4.7. The results are (not surprisingly) less marked, but the overall pattern is quite similar. The proportion of earners with 1993 earnings in what was the lower decile again rose by 2.9 points. The proportion of earners in the middle 70 per cent fell by almost 20 points, and the top two deciles increased by 3.8 and 13 percentage points respectively. There is polarisation, but again of a very asymmetric kind, with growth in the size of the top group five times that of the bottom group. This marked upwards shift may appear puzzling, but as Figures 4.1(a) and 4.1(b) indicate, the real earnings of every decile except the bottom one have increased, and the top deciles by much more than the lower. As a consequence, there has only been very limited earnings polarisation in London, although there has been a marked increase in the degree of earnings inequality.

Table 4.7 Changing proportions of earners by deciles, 1979 and 1993
(earnings adjusted to April 1993 values by median earnings change)

Deciles	Gross weekly individual earnings in 1979 £s	Size of group (%)		Percentage point change 1979–93	
		1979	1993		
London: decile cutting points					
1	66.41	10	12.9	2.9	2.9
2	116.50	10	7.7	−2.3	
3	159.11	10	7.7	−2.3	
4	188.65	10	5.8	−4.2	
5	216.93	10	6.4	−3.6	−15.9
6	243.08	10	6.9	−3.1	
7	280.04	10	7.2	−2.8	
8	320.38	10	8.6	−1.4	
9	397.39	10	13.8	3.8	
10		10	23	13	13
Britain: decile cutting points					
1	49.51	10	13.5	3.5	3.5
2	87.38	10	10.2	0.2	
3	128.15	10	9.3	−0.7	
4	161.30	10	8.9	−1.1	
5	192.23	10	8.1	−1.9	−11.4
6	222.64	10	8	−2	
7	255.4	10	7.7	−2.3	
8	296.70	10	7	−3	
9	364.10	10	9.4	−0.6	
10		10	17.9	7.9	7.9

Sources: General Household Survey, 1979 and 1993.

The polarisation of earnings, 1979–95: the NES data

It may be objected that the data used above from the GHS are based on too small a sample, or that they fail to capture earnings data. An identical exercise was therefore performed using NES data. The results are even more striking than those derived from the GHS. They show that the overall distribution of earnings was sharply upwards in Great Britain, but particularly in London. For male full-time workers the proportion of earners in the bottom decile in 1979 had declined by 1 percentage point in GB and by 6 points in London by 1993. Similar declines are found in both the bottom quartile and in the middle 50 per cent of earnings, and they were particularly marked in London. The data show remarkable increases for the top quartile in both Great Britain (23.3 points) and London (40 points). In both cases the bulk of the increase was in the top deciles. Full-time female workers showed a

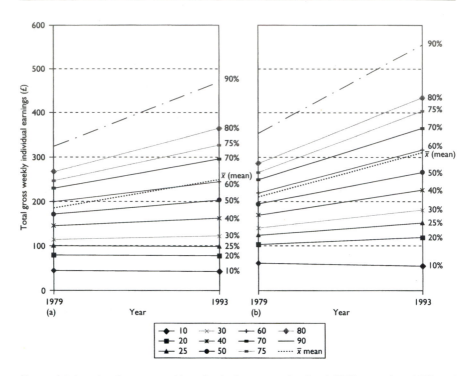

Figure 4.1 Levels of gross weekly individual earnings (at April 1995 prices) in 1979 and 1993, at deciles, quartiles and mean for (a) Great Britain (error ±15 per cent), and (b) Greater London (error ±22.4 per cent).
Sources: General Household Survey, 1979 and 1993.

similar, but even more marked, pattern. The decline at the bottom end is very sharp, particularly in London, and the growth in the size of the top group was larger than for males. There was no evidence for earnings polarisation using NES data adjusted for retail price inflation (Table 4.8). On the contrary, the data show a strong upwards shift in the earnings distribution.

Again, it could be objected that where earnings rise more rapidly than inflation across the board, as has tended to happen in Britain, then the result will be an upwards shift, which will be particularly marked if earnings in the top quartile or decile have risen much faster than other earnings, as they have. We can attempt to control for this by adjusting 1979 earnings by the median increase in earnings (Table 4.9). This shows evidence of limited earnings polarisation both in Great Britain and in London. The share of the bottom decile and bottom quartile of male full-time earnings increased in Britain in the period 1979–95 by 8 and 6 points respectively, the middle 50 per cent declined and the top quartile and top decile both increased by 6.5 and 6.9 points respectively. The results for female full-time workers were

Table 4.8 Polarisation of earnings in London and Great Britain: 1979–95 New Earnings Survey (1979 figures are adjusted for retail price inflation)

Earnings	Greater London			Great Britain		
	1979	1995	ppc	1979	1995	ppc*
Male full-time workers						
Bottom decile	10	4.0	−6.0	10	9.0	−1.0
Bottom quartile	25	10.3	−14.7	25	18.8	−6.2
Middle 50%	50	24.3	−25.7	50	32.9	−17.1
Top quartile	25	65.4	40.4	25	48.3	23.3
Top decile	10	48.7	38.7	10	30.2	20.2
Female full-time workers						
Bottom decile	10	1.7	−8.3	10	3.9	−6.1
Bottom quartile	25	4.1	−20.9	25	8.7	−16.3
Middle 50%	50	14.0	36.0	50	34.0	−16.0
Top quartile	25	81.9	56.9	25	57.3	32.3
Top decile	10	63.9	53.9	10	38.9	28.9
Female part-time workers						
Bottom decile	10	5.6	−4.4	10	5.8	−4.2
Bottom quartile	25	9.8	−15.2	25	10.1	−14.9
Middle 50%	50	32.8	−17.2	50	24.1	−25.9
Top quartile	25	65.8	65.8	25	57.4	32.4
Top decile	10	45.5	45.5	10	38.9	28.9

Sources: New Earnings Survey, 1979 and 1995.

* ppc = percentage point change.

Table 4.9 Polarisation of earnings in London and Great Britain: 1979–95 New Earnings Survey (1979 figures are adjusted by median earnings increase)

Earnings	Greater London			Great Britain		
	1979	1995	ppc	1979	1995	ppc*
Male full-time workers						
Bottom decile	10	12.5	2.5	10	18.0	8.0
Bottom quartile	25	24.1	−0.9	25	31.8	6.8
Middle 50%	50	36.8	−13.2	50	36.7	−13.3
Top quartile	25	39.1	14.1	25	31.5	6.5
Top decile	10	22.5	12.5	10	16.9	6.9
Female full-time workers						
Bottom decile	10	10.6	0.6	10	19.1	9.1
Bottom quartile	25	21.5	−3.5	25	31.6	6.6
Middle 50%	50	36.3	−13.7	50	35.6	−14.4
Top quartile	25	42.2	17.2	25	32.8	7.8
Top decile	10	21.8	11.8	10	18.1	8.1

Sources: New Earnings Survey, 1979 and 1995.

* ppc = percentage point change.

virtually identical. In London, however, the growth of the bottom decile of males and females was much smaller (2.5 and 0.6 points respectively), while growth at the top end was much larger (12.5 and 11.8).

Thus, even when using earnings-adjusted data evidence of polarisation was very limited and very asymmetrical, with a much larger growth at the top end than at the bottom. It should also be noted that adjusting the data for median earnings growth rather than retail price inflation has the effect of artificially increasing the size of the bottom end and reducing it at the top, thereby producing an impression of greater polarisation.

The growth of top earnings in London and the City

The forces driving growing earnings and income inequality in London are relatively straightforward. At the top, because of London's concentration of highly skilled and highly paid jobs in the financial, business and legal services and in some parts of the creative and cultural industries, a small but growing segment of the labour force are very highly paid both in terms of their basic salary and for some groups via bonuses on top of their basic salary. At the bottom end, the collapse of manufacturing industry and the consequent rise in unemployment since the early 1980s (notwithstanding its subsequent fall during the second half of the 1980s and the second half of the 1990s) have meant that a far higher proportion of the population are living on unemployment pay or other state benefits such as income support (London Research Centre, 1996). So, too, the growth in part-time work and low-paid service jobs has created new low-paid jobs, though whether the growth is as large as some suggest is dubious, not least because of the decline in more traditional low-paid jobs. These processes are not unique to London. They are characteristic of all global cities.

Looking at the top end of the distribution, top industrial managers are now paid very highly compared to shop-floor workers. The annual report from Income Data Services (2001) on the salary packages of FTSE 100 chief executives found that in 2000 the median package was £962,000 p.a., including salary and bonuses: an increase of 18 per cent on the previous year. The median package for FTSE 250 chief executives was about half this at £489,000. This compares to a rise of 5 per cent for all employees (Table 4.10).

Impressive as these figures are, a large proportion of the very top earners are found in financial, legal and business services such as management consultancy, investment banking, mergers and acquisitions, financial PR, advertising, and so on. To some extent, this can be seen as a reflection of the importance of finance as opposed to industrial capital within modern capitalist societies. In many respects, the movement, management and exchange of money capital has taken over from production of basic goods and

Table 4.10 Median pay levels in the FTSE 350, 2000–01

	FTSE 100 (£ p.a.)	Mid-250 (£ p.a.)
Chief executive		
Salary	538,000	300,000
Total cash	845,524	420,000
Total earnings	962,145	489,000
Finance director		
Salary	319,000	183,500
Total cash	479,955	267,494
Total earnings	371,000	304,900

Source: Income Data Services (2001).

services as the major economic activity of global cities. As Sassen (1991) argues, global cities now perform a number of key roles, one of which is the design, production and marketing of new financial instruments such as derivatives. Given the huge volumes of trades executed each day in the main financial centres, and the high turnover value involved, it only requires a small proportion of these flows to generate substantial income for companies involved. The City resembles a successful casino, not in the sense that the gains and losses are random, but in the fact that the house always wins in the long term as it takes a percentage of the stake. Second, in most financial institutions, salaries and bonuses are results-driven and comprise a very high proportion of gross profits, usually over 50 per cent. People are paid in proportion to the profits that they are able to generate through their activities, and if they are not paid sufficiently well they are likely to move to another company that will pay them more. Finance and banking are skill and knowledge businesses and remuneration for successful performers is very high, reflecting the profits they can generate. While the City has always generated high incomes, post-Big Bang the number of high earners has risen very substantially.

The City: a winner-takes-all society

In essence, the City is a classic example of what Frank and Cook (1995) have termed the 'winner-takes-all society'. In this type of society, small differences in performance between the very top performers and those who are very good or merely competent can result in very large differences in earnings. This has been intensified by growing globalisation of financial markets and their associated labour markets. Successful top traders, the 'star performers', now earn salaries running into millions of pounds a year and there are large numbers of City workers who earn well over £100,000 per year.

At entry level, Goldman Sachs paid a starting salary of £42,000 to graduate recruits in 2001, with the prospect of this rising to £80,000 within four years if they are kept on. At the top end, some have packages, including bonuses, which run into several millions. A good example was given in November 2001 when it was found that Michael Dobson, the new chief executive of Schroders, the City fund management group, was to have a guaranteed package worth £10.5 million over three years, with a guaranteed pay off of £3.5 million if he was sacked (Hughes, 2001d). The package consists of an annual salary of £200,000 plus a guaranteed bonus of £3.3 million per year. Shares worth £3.75 million will also be due if Schroders share price doubles in the next five years. Warner (2001a) also points out that in order to recruit Dobson, Schroders is purchasing his company Beaumont, a start-up fund management group with little in the way of a track record, for £33.5 million. The chairman of Schroders denied the package was excessive, arguing that 'There are other asset management firms which pay executives far more.' In December 2000 Hamilton reported that:

> Morgan Stanley, the first major institution to report, announced last week that 50 of its London staff would get a Christmas bonus of £1 million each, substantially up on last year. But the £1m bonus is now so commonplace as to be unremarkable. Rival banks poach staff with huge bonus guarantees, and one banker recently lured by a German finance house in London, according to City sources, was awarded a £6.9m guaranteed bonus over three years.
>
> (Hamilton, 2000)

Ungoed-Thomas (2000) reported that junior analysts at Morgan Stanley could expect to get bonuses of up to £50,000, the traders of up to £200,000, and the best-performing directors get bonuses of up to £4 million. Similarly, the *Financial Times* (Silverman, 2001) reported that the chief executives of three major Wall Street firms, Morgan Stanley, Merrill Lynch and Bear Sterns, each received compensation packages worth over $30 million in 2000, including salary, bonus and stock options.

Paying for specialist advice: legal and business services

The ability of financial institutions to pay these large salaries and bonuses is based on the nature of the work carried out in these sectors, the nature and wealth of the clients, the massive financial volumes involved and an ability to skim off a percentage of very large transactions. With many corporate acquisitions and mergers involving hundreds of millions of pounds, the ability of advisers to take 2–4 per cent of the total sum involved plus success fees can and does generate very large fees. A £400 million takeover could easily generate £20–40 million in fees. Similarly, in equity or foreign exchange or derivatives trading and fund management, successful firms generate many hundreds of millions of pounds in profits each year. A high proportion of this money is paid out in the form of salaries and bonuses to retain and reward the key staff involved. The closer the job to the centre of large-scale financial flows, the greater the rewards involved and the greater the salaries and bonuses paid. In Marxist parlance this is termed the ability to extract 'surplus value'. The surplus value involved in routine industrial production or service employment is relatively small, though it can be substantial when aggregated. But the scale of surplus value involved in financial and corporate transactions is far greater, as is the potential for big rewards (Coakley and Harris, 1983). Because of the dependence of many financial firms on the skills and abilities of their top staff to generate profits, bonuses, 'golden handshakes' and 'golden handcuffs' have become an accepted part of doing business and in good years add substantially to overall earnings. In financial and business services, which deal with the corporate and financial sector, partners of some of the major accountancy, advertising, public relations and marketing firms can also earn hundreds of thousands of pounds per year. In 2001 the partners at Ernst and Young, one of the top accountancy firms, each earned £449,000 per year, and partners in the top City firms typically earn between £300,000 to £500,000 p.a. (Ashworth, 2001). Ernst and Young are the government-appointed administrators for Railtrack. Their total weekly fees were around £500,000 and the job was expected to last for six months, with the lead administrator charging his time at £450 per hour and the rest of the team charging £250 per hour. The collapse of Enron, the giant American energy company, is being handled in Europe by Price Waterhouse Coopers and they have a team of fifty people with four partners, each of whom could be billing at up to £5,000 per day. It is easy to see from these examples how the senior partners can earn the sums they do.

Similar processes are at work within areas of business and legal services where major corporations dealing in many different countries require highly specialised and expert legal or management services to operate successfully. Where mergers and acquisitions or similar major ventures are concerned, the financial rewards can be huge, and so are the fees which are now charged. In major City legal firms, lawyers may be charging fees of £250 per hour, with senior partners charging £400 an hour. Top management consultancy firms typically charge out consultants at rates of £600 to £1,000 per day, with senior consultants being charged at £2,000 per day. Not surprisingly, the revenues and profits generated from such activities are considerable, as are the salaries. At the bottom end, newly qualified solicitors are being recruited at salaries of over £40,000 per year, rising to perhaps £80,000 after four years' experience.

At the top end of the earnings pyramid, senior partners in the top London law firms are estimated to be earning £1 million per year, and lower down the hierarchy other partners may be earning between £200,000 and £400,000 per year. Catrin Griffiths (2000), writing in *The Times*, noted that a group of five elite firms known as the magic circle now dominate the British legal market. They are Allen and Overy, Clifford Chance, Freshfields Brukhaus Deringer, Linklaters and Alliance, and Slaughter and May. Much of the work done by these firms is in commercial law (where the rewards are greatest), and the top London firms collectively generated £7.7 billion in fees in 2000–01. Griffiths points out that:

> They get the cream of the work, they are (mostly) more global than anyone else, and they're by far the most profitable law firms in the City. Such is their weight that the top five firms between them earned a shade under £2 billion in 1999–2000. With the numbers so high, the rivalry between these five firms has been heightened. Partner profitability – for so long a closely guarded secret – has now become a point of pride among the magic circle. At Slaughter and May, traditionally the most profitable firm in the City, 43 partners earn over £1 million a year. Small wonder then, that there was delight at Allen and Overy when 16 of their partners edged over £1 million. On the other hand, several of the most senior lawyers at blue-blooded Freshfields, who earned £810,000, were considerably chagrined this year to learn that they had only just pipped their peers at arriviste Clifford Chance where senior partners earned £805,000.
>
> (Griffiths, 2000)

Griffiths states that the five firms are at their most competitive in the global area and quotes the managing partner of Linklaters as saying that there has been 'an explosion of corporate activity around the globe, particularly in Europe, which last year saw – for the first time – a faster growth of European M&A deals compared to the US':

> An increasing proportion of the magic circle's turnover comes from abroad. At Clifford Chance, for example, the global practice outside London accounted for 37 per cent of its £585 million turnover – and that was before its mergers with US and German firms. Non-London turnover now accounts for 57 per cent of revenues.
>
> (Griffiths, 2000)

She quotes Guy Beringer, a senior partner at Allen and Overy, as saying:

> We're not any longer a UK law firm . . . The UK market is not enough in the long term to sustain law firms . . . If we don't have US capability,

we will see our traditional markets – European and Asian capital markets work – for example, being eroded.

(Griffiths, 2000)

It is not just senior partners who have benefited from the expansion of legal work and the consequent rise in profits. Salaries for assistant solicitors have also risen, and in 2001 Clifford Chance announced that it was raising pay for newly qualified solicitors from £33,000 p.a. to £42,000 p.a. Nor are these earnings restricted just to solicitors in the top City firms. A report in *The Lawyer* (Gibb, 2001b) stated that average earnings for top barristers (those lawyers who represent and argue cases in court) approached £500,000 p.a. in 2000–01. This figure refers to average gross revenues per barrister, out of which they have to pay tax and chambers overheads, and conceals large variations in individual earnings within firms; but nonetheless it shows the very high-fee income involved.

The huge salaries and bonuses in financial, legal and business services are paralleled in other areas, such as property. In July 2000 it was reported in *The Times* that Savills, the upmarket estate agent, would be paying out nearly £30 million in annual bonuses. Its ten executive directors would share £2.9 million, and branch managers of its estate agency offices could expect bonuses of up to £100,000 each (Moore, 2000). It may be objected that the cases cited are quite exceptional and restricted to a handful of senior individuals. The evidence is strongly against this. The City of London and associated financial, legal and business services commonly offer starting salaries of £25,000 to £40,000 for new graduates, and this rises rapidly with experience. High salaries are systemic. The financial services industry is highly cyclical, however, and it experiences periods of both boom and bust. The mid–late 1980s were a boom period as firms geared up for Big Bang. This was followed by a downturn and layoffs. So too, the late 1990s were a boom period which culminated in the high-tech boom in 1999–2000 with record equity volumes and large numbers of mergers and acquisitions. Fierce competition for staff ensued and salaries and bonuses reached record levels. In 2001 the bubble burst and staff are being laid off in large numbers and bonuses are being cut. In November 2001 Merrill Lynch emailed all its staff, looking for redundancies, and Woods and Godson (2001) and Pretzlik and Saigol (2001) reported that city bankers face pay cuts of 50–60 per cent, if they are fortunate enough to keep their jobs. It may be that 2001 is simply a recessionary 'blip' in a continuing long-term boom, but if it proves to be longer term the decline in City earnings may reduce inequality somewhat in the next few years.

Household income change

The trends in individual earnings in London revealed a substantial rise in inequality. It is also important, however, to examine the changes in the distribution of household income, not least because the household is the key economic unit for many purposes, including house purchase. Whereas data such as the New Earnings Survey are based on individual earnings from employment, household income data refer to all sources of income (including benefit or state transfer income) for all members of the household.

In 1975/6 average gross weekly household income in the South East was some 10 per cent higher than the UK average. By 1995/6 it had risen to 25 per cent higher than the UK average, whilst many of the old industrial regions of the Midlands and the North had average household incomes of 80–90 per cent of the UK average. Between 1979/80 and 1989/90 average household incomes increased in real terms by 37 per cent in London and 31 per cent in the rest of the South East, compared to an average of 23 per cent in the UK as a whole and just 6 per cent in Northern Ireland and 9 per cent in Wales (Stark, 1992; Buck, 1994). The data are consistent in showing that London is getting richer than the rest of the country, and has been since at least the late 1970s.

The shape of the income distribution between London and the South East on the one hand and the Midlands and the Northern regions on the other is also very different. Whereas London and the South East have a lower proportion of households in the lowest-income bands than the old industrial regions, they also have a higher proportion of incomes in the higher-income bands. In 1994, for instance, 25 per cent of households in the South East had incomes of over £550 per week, compared to 10.4 per cent in the North, 15 per cent in the North West and 8.6 per cent in Northern Ireland. These are significant differences. Comparing household income distribution for London and the UK for 1981 and 1995–8 (Table 4.11) shows that London

Table 4.11 Household income distribution, UK and London, 1981 and 1995–8

Weekly income (£s 1995–8)	1981		1995–8		1996–9	
	UK	London	UK	London	UK	London
Under £100	12	10	14	15	12	12
£100–150	11	11	11	11	10	9
£150–250	17	16	16	12	16	13
£250–350	18	16	13	13	13	12
£350–450	15	13	12	10	11	10
£450–600	14	17	13	14	14	13
£600–750	7	10	8	9	9	10
Over £750	5	8	12	17	14	21
Total	100	100	100	100	100	100

Sources: Gordon (2000) and Family Expenditure Survey, Office for National Statistics.

Table 4.12 Changing household weekly income, GB and London, 1979–93

	Great Britain			Greater London		
	1979	1993	Change	1979	1993	Change
Median £ pw	249	258	4%	269	280	4%
Mean £ pw	275	347	26%	302	392	30%
Inter-quartile ratio	3.49	4.08	0.59	3.66	5.11	1.45
Inter-decile ratio	7.4	10.10	2.70	8.09	12.27	4.18

Sources: General Household Survey, 1979 and 1993 (OPCS via MIDAS).

Note: Incomes at April 1995 prices.

had 50 per cent more of its households (21 per cent against 14 per cent) in the highest-income category than did the rest of the UK.

Analysis of raw GHS data, unadjusted for changes in household composition, shows that median household weekly income in GB and in London rose by only 4 per cent in real terms between 1979 and 1993, whereas in the rest of the South East median incomes rose by 15 per cent in real terms. The surprisingly low increases reflect changes in household composition, particularly the growth of one-person households in London which are not adjusted for and are thus somewhat misleading. Mean incomes rose more sharply than median incomes: by 26 per cent in GB and 30 per cent in London (Table 4.12).

When the figures are disaggregated and we examine mean weekly household income by decile groups, we find a very divergent pattern (Table 4.13). At the top end of the distribution incomes have risen considerably whereas at the bottom end household incomes have either risen much more slowly (in Britain as a whole) or fallen (as in London). While the lowest five deciles in Britain showed increases in real household incomes of between 2 and 7 per cent from 1979 to 1993, the top five deciles saw steadily larger increases, with the top decile seeing an increase of 55 per cent. A similar pattern was found in ROSE, but the real incomes of the bottom five deciles in London declined by between 2 and 12 per cent (the bottom decile), whereas real incomes of the top decile rose by 66 per cent. This is partly a result of changes in household composition and the growth of one-person households, but the decline at the bottom end is also likely to reflect growing unemployment and reliance on social security benefits in the 1980s and the first half of the 1990s. It should be noted here that Buck's analysis of FES data from 1978–81 to 1989–91, adjusted for changes in household size, shows increases in household incomes in London across the whole range, even for the lowest decile, although, as here, the increases are much greater at the top end.

Table 4.13 Changing household income distribution, by deciles, 1979–93

Deciles	Great Britain				Greater London			
	Mean weekly income		Change absolute	1979–93 (%)	Mean weekly income		Change absolute	1979–93 (%)
	1979	1993			1979	1993		
1	53.3	54.3	1	1.9	53.9	47.6	−6.3	−11.7
2	79.6	83.4	3.8	4.7	81.8	80	−1.8	−2.2
3	109.4	117.5	8.1	7.4	115.5	108.6	6.9	−5.9
4	160.4	163.5	3.1	1.9	172.6	157.8	−14.8	−8.6
5	220.9	224.3	3.4	1.5	238.5	228.6	−9.9	−4.1
6	272.5	296.1	23.6	8.6	297.1	320.8	23.7	8
7	323.1	377.3	54.3	16.8	351.7	421.7	70	19.9
8	380.3	474.9	94.6	24.9	422.6	544.7	122.1	28.9
9	462.1	614.2	152.1	32.9	506.5	722	215.5	42.5
10	687.7	1,069.5	381.8	55.5	776.8	1,290.1	513.2	66.1
Sample n	9,722	8,814			1,209	1,046		

Sources: General Household Survey, 1979 and 1993 (OPCS via ESRC and MIDAS).

Table 4.14 Percentage share of aggregate income, by deciles, 1979–93

Deciles	Great Britain			Greater London		
	Percentage share of aggregate income		Percentage point change 1979–93	Percentage share of aggregate income		Percentage point change 1979–93
	1979	1993		1979	1993	
1	1.94	1.56	−0.38	1.79	1.22	0.57
2	2.90	2.40	−0.50	2.71	2.03	−0.68
3	3.98	3.38	−0.60	3.85	2.78	−1.07
4	5.83	4.71	−1.12	5.72	3.99	−1.73
5	8.03	6.45	−1.58	7.91	5.85	−2.06
6	9.91	8.52	−1.39	9.77	8.21	−1.56
7	11.75	10.87	−0.88	11.67	10.69	−0.98
8	13.85	13.66	−0.19	14.02	13.93	−0.09
9	16.80	17.69	0.89	16.80	18.30	1.50
10	25.01	30.76	5.75	25.78	33.00	7.22
Sample n	9,722	8,814		1,209	1,046	

Sources: General Household Survey, 1979 and 1993 (OPCS via ESRC and MIDAS).

Table 4.14 shows each decile's changing share of aggregate income. It is clear from this that not only do the bottom deciles have a very small share of total household income (under 2 per cent for the bottom decile) compared to the top one (over 25 per cent for the top decile), but the disparities have widened considerably over time, particularly in London. In 1979, the bottom decile accounted for 1.79 per cent of total household income in London; by 1993 this had fallen to 1.22 per cent. The shares of all deciles up to and including the eighth showed a fall, but the top decile increased its share of total income by a remarkable 7.22 percentage points: from 25.8 to 33 per cent. In other words, the income share of the top 10 per cent of all households increased from a quarter to one-third of total household income. The same pattern can be seen in Britain as a whole, but the bottom deciles have a slightly greater share of total income, and the increase in the share of the top two deciles is less than London. In GB the share of the top decile rose from 25 per cent to 30.8 per cent. Work by Mollenkopf (1998) and Preteceille (2001) show similar trends for both New York and Paris. In both cities, the top decile of households took an increasing share of total household income during the 1990s.

The inter-quartile and inter-decile ratios summarise the changing extent of income inequality (Table 4.12). In Britain, the inter-quartile ratio rose from 3.49 in 1979 to 4.08 in 1993 and the inter-decile ratio rose from 7.4 to 10.1. In other words, in 1993 the top decile incomes were ten times those of the lowest decile. In London, the inter-quartile ratio rose from 3.66 to 5.11 and the inter-decile ratio rose from 8.1 to 12.3. In a nutshell, income inequality was higher in London than in Britain as a whole in 1979, and it has increased further over the period. It is important to note that the inter-decile ratios are much higher for incomes than for earnings. This is because household incomes run from households with those with two or more professional high earners to those with no earners at all who are solely dependent on state welfare benefits.

It should be noted here that Buck's (1997) analysis of changes in household incomes in London and the UK, which used pooled FES data for the periods 1978–81 and 1989–91, reaches broadly similar conclusions. He notes 'an enormous increase in inequality in the UK, and an even more substantial increase in London. In the late 1970s the inter-decile ratio in London was only marginally greater than that for the UK as a whole. By the early 1990s it was nearly 40% greater' (p. 13). But when he divided household income change into those households where the principal earner was employed, and those for where the household head was unemployed, he found that the growth in the level of inequality for households with employed heads was similar in London to that of the UK as a whole. He suggests that this points to national explanations of change, and when he decomposes changes in income inequality he found that a large amount of the increase in inequality in London can be accounted for by both a higher growth of unemployment in London and by the higher and more rapidly rising incomes of those in work. He concludes that while the occupation effect is somewhat higher in London than in the UK as a whole, 'it is not apparently change within the employed population that is making London distinctive, but changing patterns of employment and the changing position of the unemployed relative to those in employment' (p. 15). More specifically, he points to the fact that by 1996 almost a quarter of Inner London residents in working-age households had no one in the household with a job. Overall, he points to the greater proportion of households at risk of unemployment in London, including single-person households, lone parents and ethnic minorities.

The polarisation of household incomes?

We know that high-income households are considerably better off in real terms than in 1979 and that the gap between them and low-income households has increased. But to what extent, if at all, has the proportion of high- (and low-) income households grown as a share of the total? By comparing quartile and decile groups in 1979 and their size in 1993 (adjusting the income-cutting points for inflation) the answer is unambiguous. In Britain as a whole, the proportion of households in the bottom income band (1979 lower quartile adjusted for retail prices) fell by 2.5 points and the proportion in the middle band also fell by 7.3 points. By comparison, the proportion in the top band rose by 9.8 points. In London, the bottom band grew by 2.2 points and the top band increased by 10.3 points (from 25 to 35.3 per cent). Decile analysis shows that the share of households in the top income band in 1993 (top band in 1979 adjusted for inflation) grew from 10 per cent to 23.6 per cent. The number of households in the top income band has risen sharply in London. Similarly, we find the proportion of households with incomes of less than half median income in London rose 0.5 points from 31.8 to 32.3 per cent (almost a third of households) between 1979 and 1993, whereas the proportion with over twice median income rose by no less than 16.3 points from 7.6 per cent to 23.9 per cent.

The impact of changes in the distribution of household types and number of earners

One of the key drivers of the increase in household income inequality has been the changing distribution of household types and the number of income earners. Table 4.15 shows that the number of households with no earners in London rose from 30 per cent in 1979 to 41.4 per cent (the rise in Great Britain was almost identical). The number of one-earner households remained stable at 34 per cent, while the number of two- and three-earner households fell from 36 per cent to 24.2 per cent. Clearly, where an increasing proportion of households have no earners and are dependent on state benefits, which are both lower than median earnings and have tended to rise in line with retail prices rather than earnings, the income gap between households with no earners and those with one or more earners is likely to

Table 4.15 The changing distribution of households by number of earners, Greater London and Great Britain, 1979–93

	Greater London (%)			Great Britain (%)		
	1979	1993	Change	1979	1993	Change
No earners	30	41.4	11.4	31.2	41.1	10
One earner	34	34.4	0.4	31	28.4	−2.7
Two earners	28.4	20.6	−7.8	30.2	26.2	−4
Three + earners	7.6	3.6	−4	7.6	4.3	−3.3
All households	100	100		100	100	

Sources: General Household Survey, 1979 and 1993.

Table 4.16 Absolute and percentage change in median gross weekly household income by number of earners, Greater London and Great Britain, 1979–93 (£1995)

	Greater London			Great Britain		
	Median income		Change (%) 1979–93	Median income		Change (%) 1979–93
	1979	1993		1979	1993	
No earners	84	105	25.3	83	111	33.0
One earner	258	355	37.6	242	302	24.8
Two earners	409	657	60.5	368	487	32.4
Three + earners	608	856	40.9	538	651	21.1
All households	269	280	3.9	249	258	3.6

Sources: General Household Survey, 1979 and 1993.

increase. This is seen in Table 4.16 which shows that the median incomes of households in London with no earners rose by 25.3 per cent in real terms, while the incomes of households with one earner rose by 37.6 per cent, and those with two earners rose by a remarkable 60.5 per cent. The seeming paradox that median income for all households rose by only 3.9 per cent over the period 1979–93 is explicable by the dramatic increase in the proportion of no-earner households.

The pattern is similar for mean incomes. Two-earner households in London saw the greatest increase in real incomes (+77 per cent). In cash terms no-earner households saw their mean real incomes rise by £50 per week, while two-earner households had a rise of £334. When we compare no-earner household to dual-earner professional and managerial households (not shown), the disparity is even greater. The median gross income of double professional and managerial earner households in 1993 was nearly

£1,000 per week, well above the £657 of all two-earner households and four times the median for all households in London. Although relatively few in number, two-earner professional and managerial households constitute a key part of London's new affluent and are the opposite pole to the no-earner households dependent on state benefits.

Summary and conclusions

The conventional wisdom in much of the literature on global cities is that they have experienced an increase in earnings and income polarisation as a result of the changes which have taken place in industrial and occupational structure. The growth in highly skilled and highly paid financial and business service occupations has been matched by an increase in low-skilled and low-paid personal service sector jobs, combined with work in manufacturing sweatshops. The occupational, earnings and income data for London do not support this thesis. Rather, London (like Amsterdam and Paris) has been characterised by an increase in the number and proportion of highly skilled and highly paid professional, managerial and technical workers in the service sector but by a decline in the number of less-skilled workers.

The growth of the professional and managerial group has been linked to a marked increase in earnings of this group, particularly those working in the City of London. London has seen a major upwards shift in earnings which has dramatically increased the number of high earners and the level of their earnings. As a consequence, London has been characterised by a sharp increase in inequality between rich and poor. But, and this is crucial, it does not appear to have seen a marked increase in the relative size of the poor group which the polarisation thesis predicts. Any increase in polarisation (using median incomes as the adjustment rather than retail prices) has been highly asymmetric, with a far bigger increase at the top end of the distribution than the bottom. London has become a far more unequal city, and at the top end the double-earner professional and managerial households form a core group of the well-off and pose a sharp contrast to the large group of no-earner households at the bottom.

It is important to make one further point. The growth of inequality in London and in other world cities such as New York, Paris and Amsterdam is not an inadvertent or accidental by-product of their leading global role in financial, business and creative services or their peculiar industrial structure. It is a direct consequence of this role. Earnings inequality is an inscribed characteristic which is inevitably associated with their role and occupational structure. To a very significant extent, the economy of the City of London and the salaries which are paid there generate massive inequalities.

Chapter 5

Multi-ethnic London

Migration, race and ethnicity

Introduction

Because of their key role in the international economy, and the size and structure of their labour markets, world cities have long functioned as major destinations for both national and international migration (Friedmann and Wolff, 1982). Like magnets, they attract the ambitious, the hopeful and the desperate. The role of big cities as major migrant destinations is not new; they have always had diverse populations. During the seventeenth century, the population increase in London was fuelled by large-scale in-migration, and Wrigley (1967) showed that almost one in four of London's population were immigrants. London's population also grew dramatically during the nineteenth century from one million in 1801 to 6.5 million in 1901, and a high proportion of this growth was accounted for by immigration, primarily from other parts of Britain but also from Ireland and continental Europe. The Irish in the nineteenth century can be seen as the first large-scale economic migrants to Britain, and the Huguenots in the eighteenth century and the Polish and Russian Jews in the late nineteenth century and the early twentieth were some of the first political refugees and asylum seekers. Similar migration patterns characterise nineteenth-century Paris (Rhein, 1996), which sucked in migrants from all over provincial France, and the American cities of the nineteenth and early twentieth centuries which were primarily peopled by overseas immigration (Ward, 1989). Engels's (1846) depiction of nineteenth-century Irish migration to Britain in the aftermath of the Irish famine in his *The Condition of the Working Class in England* could apply to many similar groups today:

the rapid extension of English industry could not have taken place if England had not possessed in the numerous and impoverished population of Ireland a reserve at command. The Irish had nothing to lose at home, and much to gain in England; and from the time when it became known in Ireland that the east side of the St George's Channel offered steady work and good pay for strong arms, every year has brought armies of the Irish hither. It has been calculated that more than a million have already immigrated, and not far from fifty thousand come every year, nearly all of whom enter the industrial districts, especially the great cities, and there form the lowest class of the population. Thus there are in London, 120,000; in Manchester, 40,000 . . . poor Irish people.

(Engels, 1969, p. 122)

Although the second half of the nineteenth century and the early years of the twentieth saw large-scale migration from Europe to the United States, the scale of migration has greatly increased in range and diversity in the last forty years with the rise of air travel and the growth of globalisation, and Castles and Miller (1993) have termed the second half of the twentieth century the Age of Migration. But, as in the nineteenth century, migrant destinations tend to be highly focused, and Clark (1996) has identified a number of key gateway cities for migration in the USA. London is one of the world's major global cities, with economic and financial ties to many parts of the globe. It also had long-standing colonial ties to a variety of nations. Given this role and history, and the propensity of migrants, both skilled and unskilled, to be attracted to major cities, it is not surprising that London is a destination for a variety of migrants – legal and illegal, skilled and unskilled – from all over the world.

Migrants come to big cities for a variety of reasons, ranging from job opportunities or a career, the cultural and social life associated with big cities, to those who are fleeing persecution or seeking asylum. Some also come to study or to join family members already here. But the dominant reason is for work: whether it be a well-paid job in the financial institutions, business or creative industries, low-paid work in the many bars, restaurants and hotels which service residents, tourists and the transnational elite or in the low-paid cleaning or security sectors, or even in the informal 'cash in hand' economy such as carwash workers or street sellers where pay is well below the minimum wage. The working and living conditions of most migrants have greatly improved in the last 150 years, but the underlying principle remains essentially the same. Because of their limited education, poor English, a need for money and their desire to get on, low-skill migrants frequently find themselves being exploited as cheap labour in one form or another. This was true of Jewish migrants to the East End at the end of the nineteenth century, and it is true of Bangladeshi migrants of the last twenty years, both of whom found niches in the clothing and market trades. Today Chinese and Albanian migrants can be seen working in the kitchens of Soho restaurants or in London car washes where no questions are asked about legality.

Three main waves of post-war migration into Europe have been identified by Fielding (1993). The first wave, which started in the 1950s, was labour migration into the boom economies which faced severe labour shortages; the second, which marked the 1970s and 1980s, was family reunification, as the first wave of predominately male migrants brought over their wives and families. The third wave, which Fielding labels the 'post industrial', comprises an unusual mixture of refugees and asylum seekers and skilled international migrants. Although migrants offer a ready source of low-wage labour, the potential flows are very large, and one of the major problems in Britain and other western European countries in the last decades of the twentieth century has been to try to restrict the flow of low-skill labour migrants, refugees and asylum seekers from less-developed countries while encouraging high-skill migration. This accounts for New Labour's attempts to tighten up on refugees and asylum seekers and the introduction of the Highly Skilled Migration Programme in December 2001 (Ford, 2001) and the EU's attempts to formulate a common European migration policy.

International migration to London

In the post-war period, London and other major British cities have seen successive waves of in-migration, from Ireland, southern Europe, and the New Commonwealth in the 1960s, but more recently from further afield. Initially, most migrants were relatively low skilled, doing the dirty, low-paid or anti-social hours jobs which the host population no longer wanted to do. They fulfilled the role identified by Marx of a low-wage 'reserve army of labour' and they came from countries where wages were low, unemployment high and living conditions poor. In Britain, the immigrants came from Ireland, the Caribbean, India and Pakistan, in France from Algeria and French West African colonies, in the Netherlands from Surinam and Indonesia. To an extent, the geography of European colonisation was reflected in the early pattern of post-war immigrant flows. Immigrants were recruited from a number of ex-colonial countries and, where this was not possible, in Germany, Austria and Switzerland migrants were recruited in Greece, Turkey and Yugoslavia (King, 1993; Gould and Findlay, 1994). In recent years, however, increasing globalisation has resulted in a growing number of refugees and asylum seekers from around the world. Britain has become one of the main European destinations for refugees and asylum seekers in the 1990s, attracting some 70,000 to 90,000 a year. In 2001, the UK received a total of 88,300 applications for asylum, the same as in Germany, but well ahead of France (53,875), the Netherlands (32,580) and the other EU countries. Italy, for example, had only 9,620 applications (Home Office, 2002). London receives a substantial proportion of this flow. From 1991 to 1998 it received a total of 210,000 asylum seekers or visitor switchers assigned to London – an average of 30,000 per year (Heath and Hill, 2002) and about 30 per cent of the UK total. In addition, the increasingly integrated nature of western economies has led to the growth of skilled international migration, particularly in the financial and business services sectors where the large number of overseas banks and companies has resulted in large inflows of overseas staff

(Beaverstock, 1991, 1994; Beaverstock and Smith, 1996). Jenkins, in an article entitled 'Thank God for the City's foreigners', argues that this has been very beneficial for the City, and for London as a whole:

> The iron law of successful cities is, always welcome a newcomer. In the 13th century, London banking was dominated by Italians, hence Lombard Street. When London's first great merchant, Sir John Gresham, pleaded with Elizabeth I to relax immigration for Flemish refugees, his argument was crisp: 'it will profitt the citie'. It did. In the 18th and 19th centuries, London played host to the trading houses of Europe. What we regard as British banking names were all once foreign immigrants: Rothschild, Baring, Hambro, Montagu, Schroder, Kleinwort, Warburg.
> (Jenkins, 1999)

He adds, with a reference to the words inscribed on New York's Statue of Liberty, that 'If I were running the modern City, I would erect the same statue at Heathrow airport. The City's future lies not on the playing fields of Eton but in the airport arrival lounge.'

The key role of international migration to London has been comprehensively analysed by Dobson and McLaughlin (2001) (Table 5.1). Using the International Passenger Survey, work permit statistics and Home Office data on asylum seekers they found that while the level of migration outflows has remained more or less constant from 1975 to 2000, the level of inflows has increased, and at the end of the 1990s the net inflow to the UK averaged 100,000 per year. Of the migrant inflows to the UK, London consistently receives a third, and the rest of the South East about 17 per cent. London and the South East together therefore account for half of all migrant inflows to the UK. Of the outflows, just over a quarter came from London over the period as a whole. As a result, London has received a net inflow of around 450,000 recorded migrants from 1975 to 2000, and this figure increased

Table 5.1 International migration, 1975–99, London and the UK, net flows

	All citizenships		British citizens		Non-British citizens		London as % of UK
	UK	London	UK	London	UK	London	
1975–79	−105.5	41.7	−305.2	−23.4	199.6	65.2	32.7
1980–84	−136.1	39.6	−308.1	−24.8	172.3	64.4	37.4
1985–89	121.1	115.7	−113.9	−5.1	234.9	120.8	51.4
1990–94	112.4	40.9	−109.1	−15.9	221.3	56.9	25.7
1995–99	412.3	215.7	−95.7	7.3	400.1	160.6	40.1
1975–99	404.3	453.6	−931.9	−61.8	1,228.1	467.9	38.1

Source: Dobson and McLaughlin (2001), based on the International Passenger Survey.

dramatically from 41,000 in the period 1990–4 to 216,000 in the period 1995–9. There was a similar upsurge in the second half of the 1980s from the first half of the decade, which strongly suggests that higher net inflows are a response to economic boom periods. Of non-British citizens, London accounted for just under 40 per cent of the net inflow over the period 1975–99: a total of about 468,000 people. International migration into London has offset the loss of a substantial proportion of residents through net migration within the UK. Gordon (2000) points out that the total gain in London's population of 232,000 from 1991 to 1997 was made up of a natural increase of 229,000, a loss of 283,000 through net migration to the rest of the UK, and a gain of 178,000 asylum seekers and visitor switches and 89,000 from other international migration. As we shall see later in the chapter, these migration balances, combined with differential ethnic minority age structure and birth rates, are changing the ethnic composition of London's population quite rapidly. Ethnic minorities are also far less likely to migrate out of London than the indigenous white population.

Migrants to Britain are concentrated in the 15–34 age group which accounts for just over 60 per cent of all immigrants. Students form a substantial proportion of migrants but employed migrants account for just over half the total of all migrants. Dobson and McLaughlin found that, contrary to the academic consensus about the low-skill nature of contemporary labour migrants, proportions of managerial and professional workers had remained steady at around 60 per cent of the total employed migrants from 1980 onward with clerical and manual workers steady at 40 per cent of the total. This does not suggest a vast army of low-wage workers entering the personal service sector in Britain as some would argue. In terms of employed non-British migrants, the largest proportion (28 per cent) have come from the Old Commonwealth (Australia, Canada, New Zealand and South Africa), followed by EU/EFTA (26 per cent), the rest of the world (19 per cent), other foreign developed countries (16 per cent), Indian sub-continent (8 per cent), and Eastern and other Europe (4 per cent). The researchers point out that Old Commonwealth and EU/EFTA citizens formed the two largest components of the inflow, together comprising well over half the total. When migrants from Eastern and other Europe (including Albania, Bulgaria, Turkey and Romania) and other foreign developed countries (USA, Israel, Japan, Singapore, Hong Kong, Taiwan) are added, migrants from the developed countries account for about 75 per cent of total inflow from the mid-1980s onwards, rising to 80 per cent in the second half of the 1990s. Obviously, these figures cannot and do not include illegal migrants who evade or avoid the International Passenger Survey, and these groups are likely to come from south-eastern Europe and the less-developed world, but the figures show the growing importance of migration from developed countries over the last twenty-five years. They provide strong support for Fielding's notion of a post-industrial migration flow.

London has also experienced consistent in-migration from the rest of Britain, though these flows have been counterbalanced by outflows (Ford and Champion, 2000). These inflows have been associated with the role of London as an 'escalator region' which offers substantial upwards social mobility, particularly to young graduates, professionals and other white-collar workers (Fielding, 1989, 1993) who come to London for university education, or on graduation, and then progress up the career ladder which the wide range of jobs and occupations in London offers. The upwardly mobile middle classes may then move out of London in their thirties and forties to look for a family house further out. As a result, the age structure of internal migrant inflows into London is younger than that of the outflows.

As a result of the migration streams from the former colonies and elsewhere, London, like other major world cities, is characterised by a high level of ethnic diversity with major groups from many nations and backgrounds, in addition to a large British-born ethnic minority population. But, while large, they do not bear comparison with the large cities in the United States, many of which are 'majority minority' cities (Mollenkopf and Castells, 1991; Logan, 2000; Clark, 1999a; Waldinger, 1996).

The geographical distribution of ethnic minorities in Britain

It is important before discussing the ethnic minority population in London to define what is meant by this term. Until the 1991 census, the only systematic data available on what Peach terms 'racialised minorities' had been on birthplace or parental birthplace, and the official category was of persons born in the New Commonwealth (e.g. the Caribbean, India and Pakistan, etc.) as opposed to the Old Commonwealth of Australia, Canada, South Africa and New Zealand. The 1991 census introduced a question on 'ethnic identity' in which respondents classified themselves in terms of a pre-defined set of categories. This permitted for the first time an accurate assessment of the size of the substantial second-generation ethnic minority population; that is, those people born in Britain of non-European white ethnic origin.

When the first waves of New Commonwealth migrants came to Britain in the 1950s, they moved overwhelmingly to a relatively small number of large cities: London, Birmingham, Manchester, Leeds and Bradford. This was for the simple reason that this was where the job opportunities were and, in the case of London, London Transport and the NHS were recruiting workers. The big northern cities offered jobs in the textiles industry, particularly for those willing to work night shifts. Subsequent migrants followed their earlier compatriots, partly because of the social networks which were then in place (Hamnett and Randolph, 1988b; Ramdin, 1987; Sarre, 1989).

To a large extent, the contemporary geography of ethnic minority residence in Britain has remained broadly the same as it was twenty or thirty years

Table 5.2 London's percentage share of ethnic minority groups in the UK, 2000

	Inner London	Outer London	London
Black Caribbean	29.9	23.8	53.7
Black African	48.1	30.3	78.4
Black Other	34.4	21.6	56
Indian	7.1	34.8	42
Pakistani	6.3	16.6	22.8
Bangladeshi	47.4	9	56.5
Chinese	22.4	23.2	45.6
Other Asian	21.3	39.6	60.9
Other others	29.8	25.6	55.4
All minority groups	23.1	26	49.1
White	3.4	6.4	9.8
All ethnic minority groups	4.8	7.7	12.6

Source: Scott *et al.* (2001).

ago. Ethnic minority groups are overwhelmingly concentrated in the big industrial cities and in some of the smaller industrial cities of the Midlands such as Leicester. In 1991 some 76 per cent of the total ethnic minority population of Great Britain were concentrated in the major conurbations: 45 per cent of minorities were located in London, 12.4 per cent in the West Midlands conurbation, 5.4 per cent in West Yorkshire, 4.9 per cent in Greater Manchester, and 6.2 per cent in the East Midlands, with under 1 per cent each in Merseyside and Tyne and Wear (Green and Owen, 1995; Peach, 1996). Estimates provided by Scott *et al.* (2001) show that the concentration of ethnic minorities in London had increased slightly by 2000 to 49 per cent of all minorities in Britain. The proportions of different ethnic groups in Britain who live in London all range around the 40–60 per cent level, with the exception of Black Africans, 78 per cent of whom live in London (see Daley, 1998), and Pakistanis, only 23 per cent of whom live in London. By contrast, only 10 per cent of the white population of Great Britain live in London, which highlights the concentration of ethnic minority groups in London compared to the rest of Britain (Table 5.2). Although there are now substantial minority concentrations in a number of the smaller industrial towns and cities of the South East such as Slough, High Wycombe, Oxford, Luton, Bedford and the Medway ports, the overwhelming concentration still remains in Greater London.

The changing size and composition of London's ethnic minority population

The ethnic composition of London has changed dramatically in the last forty years. In the 1960s, the population of London was still predominantly white and British born. Although the West Indians were already established in Notting Hill and Brixton, the Irish were a major group in Camden Town, Kilburn, Fulham and other parts of Inner London, and Greek Cypriots were established in Camden Town and Haringey, other groups were still relatively thin on the ground. Although data on ethnic origins did not become available until the 1991 census, the 1971 census shows the proportion of residents of Greater London born in the New Commonwealth (the Caribbean, India, Pakistan, Sri Lanka, Bangladesh, and a number of African countries such as Nigeria, Ghana, Kenya and Zimbabwe) was 6.4 per cent and the proportion with both parents born in the New Commonwealth was 7.6 per cent. In Inner London, the proportions were not much greater at 8 per cent and 11 per cent respectively (see Deakin, 1974). This is distinguished from the Old Commonwealth consisting of South Africa, Canada, Australia and New Zealand who had all gained their independence from Britain much earlier.

In the subsequent thirty years, the ethnic minority population has grown substantially in both size and variety and London has shifted from being primarily mono-ethnic to being multi-ethnic. Storkey and Lewis (1996) term it a 'true cosmopolis'. In Haringey, which now has a minority non-white population totalling 34 per cent, there is a substantial Greek and Turkish Cypriot population who came to Britain from the 1950s onwards, a South Asian population complete with Halal butchers and Sari shops, and a large West Indian population in the east and south of the borough. There is also a large sprinkling of Somali, Ethiopian, central African, Bosnian and Kosovan refugees. Unsurprisingly, the council print some of its community leaflets in seven or eight languages.

The scale of the transformation is remarkable. By 1991 the ethnic minority population of Greater London had reached 20 per cent and 25 per cent in Inner London. The 2001 census showed that a third (34 per cent) of the population of Inner London and 28 per cent of the population of Greater London consisted of ethnic minorities, an increase of 53 per cent on 1991. Table 5.3 shows that the various black groups comprised 11 per cent of the population of Greater London and 17 per cent of the population of Inner London. The Bangladeshis are also strongly concentrated in Inner London with 4.4 per cent of the population compared to just 0.5 per cent in Outer London. But by contrast, the Indian and Pakistani ethnic groups are concentrated in Outer London, where they comprise 10 per cent of the population against 4 per cent in Inner London. The Chinese and other Asian groups are equally distributed with 3 per cent in Inner and Outer London.

The ethnic minority population of London has grown very rapidly in recent decades. This is partly a result of the differences in the age and fertility structure of the ethnic minority groups compared to the host population rather than new migration, although this has been important for some groups. Scott *et al.* (2001) estimate that the Chinese population of Britain grew by just 2 per cent between 1992–4 and 2000, the Black Caribbean

Table 5.3 Population estimates (in 000s) for London by ethnic group, 2000

	Inner London	Inner (%)	Outer London	Outer (%)	London	London (%)
Black Caribbean	158	5.8	126	2.8	284	4.0
Black African	212	7.7	133	3.0	345	4.8
Black other	105	3.8	66	1.5	171	2.4
Indian	70	2.6	343	7.8	412	5.8
Pakistani	42	1.5	112	2.5	154	2.1
Bangladeshi	122	4.4	23*	0.5*	145	2.0
Chinese	33	1.2	35	0.8	68	0.9
Other Asian	51	1.9	96	2.2	147	2.1
Other others	137	5.0	117	2.6	254	3.5
All minority groups	931	33.9	1,051	23.8	1,982	27.7
White	1,810	66.0	3,368	76.2	5,177	72.3
All ethnic groups	2,743	100.0	4,422	100.0	7,165	100.0

Source: Scott *et al.* (2001).

* small numbers.

population grew by 4 per cent, the Indian population by 9.3 per cent, the Pakistani population by 26 per cent, the Bangladeshi by 40 per cent, the Black other group by 61 per cent, and the Black African group by 67 per cent from 263,000 to 440,000. These figures reflect the sharp differences in age structure of the respective ethnic groups. The median age of the White population is 37 years, and 31 years for both Chinese and Indians, compared to 18 years for both the Black other and Bangladeshi populations (Figure 5.1). The age structure of these groups has implications for future population growth. Scott *et al.* (2001) also point out that the age and sex profiles and the proportion of each ethnic group born in the UK were influenced by the timing of the various waves of immigration into Britain.

> The first large scale migration of immigrants of ethnic minority origin came from the Caribbean shortly after the Second World War and during the 1950s. Those from the Caribbean generally came as families compared with the second main group of migrants from India and Pakistan, who tended to arrive as male adults with their wives and children following. Many immigrants from these groups arrived before the 1971 Immigration Act came into effect. Most Chinese people emigrated to Britain in the 1980s, as mainland China opened . . . Immigration from Bangladesh peaked in the early 1980s and many of the Black Africans in Britain came in the 1980s and 1990s.
>
> (Scott *et al.*, 2001, p. 11)

There are also major differences in fertility between various groups. The fertility rate for Pakistani and Bangladeshi women is much higher than the

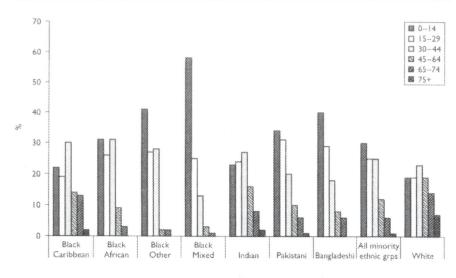

Figure 5.1 Age distribution by ethnic group, Great Britain, 1997–9.
Source: Scott *et al.* (2001).

rate for other groups. A remarkable 40 per cent of the Bangladeshi population of Britain, and 41 per cent of the Black other group, is aged under 15,
compared to 19 per cent for the White population. Scott *et al.* (2001) note:
'Their young age structure and the consequential large number of births
and relatively small number of deaths helps to explain the disproportionate
contribution of minority ethnic groups to population growth in the 1990s'
(p. 11). They point out that 'Virtually all (97 per cent) persons aged over 44
from all the minority ethnic groups were born outside the UK. In contrast,
90 per cent of minority ethnic children aged under 15 were born in the UK.'
The implications of these findings are that the ethnic minority population of
Britain and the major cities is likely to continue to grow rapidly over the
next ten to twenty years. It is likely that the ethnic minority population of
Inner London will exceed 40 per cent by 2010 at the current rate of growth.
In some boroughs, such as Brent and Newham, the proportion already
exceeds 50 per cent, and in Tower Hamlets, Hackney, Ealing and Harrow
the proportion exceeds 40 per cent. The effect of the differential age structure of ethnic minority groups is shown by the ethnic composition of pupils
in state schools. In Greater London as a whole children of white ethnic
background comprised 60 per cent of secondary school children in 1999, but
in Inner London the figure was just over 40 per cent. The proportion of
children from ethnic minority backgrounds ranged from under 10 per cent
in outer suburban boroughs such as Bexley and Bromley to 70 per cent in
Tower Hamlets and Newham and 80 per cent in Brent (DfES, 1999) (see
Figures 5.2, 5.3).

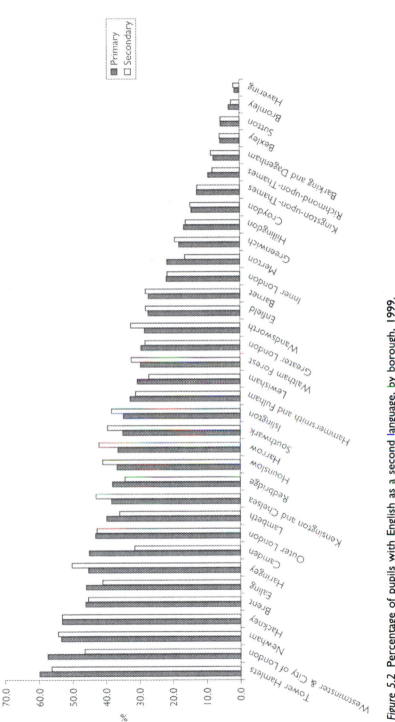

Figure 5.2 Percentage of pupils with English as a second language, by borough, 1999.
Source: Department for Education and Science, *Ethnic Minority Pupils and Pupils for whom English is an Additional Language* (London, 1999).

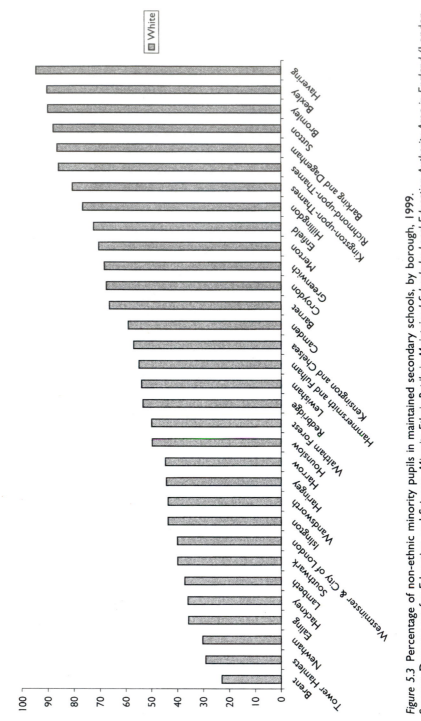

Figure 5.3 Percentage of non-ethnic minority pupils in maintained secondary schools, by borough, 1999.
Source: Department for Education and Science, *Minority Ethnic Pupils in Maintained Schools by Local Education Authority Area in England* (London, 1999).

Ethnic minorities and the labour market

Just as most of the Irish who migrated to Britain in the nineteenth century came to fill relatively menial or unskilled jobs such as building labourers, many of the first New Commonwealth migrants to Britain filled similar jobs. West Indians were recruited by the National Health Service and by London Transport in the 1960s, and the Pakistanis and Indians initially tended to move into routine manufacturing or service sector jobs. Similarly, after their arrival in Britain in 1972, many middle-class Ugandan Asians who had a commercial background moved into small neighbourhood retail businesses:

> The post-war migration to Britain from the Caribbean and the Asian sub-continent, while based upon imperial ties, was very much driven by economic imperatives. The rebuilding of the war-shattered economy created a demand for labour that could not be satisfied by the British population alone. The demand was particularly acute in the National Health Service, in public transport and in many sectors of manufacturing; qualified and unqualified labour, especially young single men, were invited from the Caribbean and the sub-continent to fill the vacancies.
>
> (Modood, 1997, p. 83)

These early positions have exercised a considerable influence on their subsequent labour market careers, and minority groups were particularly hard hit by the decline of manufacturing from the late 1960s and 1970s onwards (Hamnett and Randolph, 1988b; Cross, 1992; Cross and Waldinger, 1992). 'Ethnic minority groups in Great Britain have fared relatively badly in the labour market, tending to be employed in low-status and poorly paid jobs in declining industrial sectors' (Owen and Green, 1992, p. 7). They have also been susceptible to high levels of unemployment. In 1991 in the early stages of the last recession, unemployment among Black groups and Bangladeshis was double or treble the level among the White population (Green and Owen, 1995).

It would be a mistake, however, to assume that the labour market positions of ethnic minority groups have been generally uniformly poor. A growing divide has opened up between the labour market positions and earnings of different minority groups (Peach, 1996). Many second-generation minorities from Indian, Ugandan Asian and Chinese backgrounds have done well at school, and have gone on into successful managerial and professional careers in pharmacy, accountancy, medicine, banking and finance. By contrast, those from West Indian and African ethnic backgrounds have generally done less well at school, a much higher proportion are unemployed, and those in work have tended to be concentrated into less-skilled and lower-paid jobs. This has also been true of those of Bangladeshi ethnic origin, but it seems a

number of second-generation Bangladeshis are now progressing through the
school system and moving into higher education and hence into white-collar
jobs. As Modood comments:

> In the 1970s and 1980s theorists sought to explain racial inequality:
> what needs to be explained is racial inequality *and* ethnic diversity. For
> insofar as there is fundamental divide in employment by ethnicity, it is
> not a black–white divide, but a divide between Whites, Chinese, African-
> Asians on the one hand, and Bangladeshis and Pakistanis on the other,
> with Indians and Caribbeans in perhaps an intermediate position.
>
> (Modood, 1997, p. 84)

Not only are unemployment rates higher for some ethnic minorities, but
their pattern of employment by occupational status also differs considerably
from that for the white population. Figure 5.4 shows the socio-economic
distribution of different ethnic groups in London in 1991, ranked from left
to right in terms of the proportion of professional and managerial workers
in each group. The proportion of professionals and managers varies from a
high of 35 per cent among those of Chinese ethnic origin, and 26 per cent
for whites and 25 per cent for Asians, to lows of 16 per cent for those of
Black African origin and just 10 per cent for those of Caribbean origin.
Conversely the Chinese had the lowest proportion of skilled manual workers

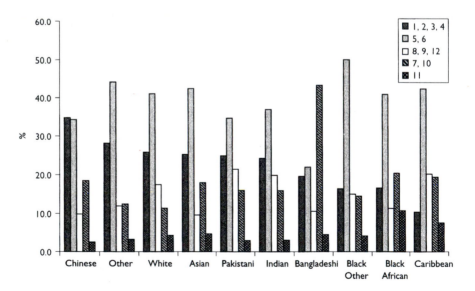

Figure 5.4 Socio-economic groups by ethnicity, Greater London, 1991.
Source: Census of Population, 1991.

Table 5.4 Income distribution by ethnic group, London, 1996

Quintile	Bangladeshi	Pakistani	Black African	Black Caribbean	Black Other	Chinese	Indian	All non-White	White
I (lowest)	76	36	35	30	31	27	18	28	13
2	10	24	20	20	18	14	23	20	14
3	10	20	23	23	21	16	26	23	20
4	1	13	15	20	21	23	22	19	26
5 (highest)	2	7	7	7	10	20	11	10	27

Source: Anderson and Flatley (1997).

(10 per cent) compared to 20 per cent for those of Pakistani, Indian and Black Caribbean origin. The percentage of semi-skilled workers was highest among Chinese (probably restaurant workers), Black African and Black Caribbean (both around 20 per cent), and the proportion of the unskilled was highest amongst Black Africans (11 per cent) compared to around 3 per cent for Chinese, Indians and Pakistanis. It is very clear that the occupational distribution of ethnic groups in London differs considerably, with the Chinese being bipolar in their distribution between managers and professionals on the one hand and semi-skilled workers on the other, whereas Black Africans and Black Caribbeans are very under-represented in managerial and professional jobs.

Not surprisingly, the differences in employment and occupational structures between different groups are reflected in the distribution of average incomes. In 1994, a national survey for the Policy Studies Institute found that white male full-time employees earned £336 per week on average, similar to the Chinese (£336) and African Asians (£335). This was followed by Caribbean men (£306), Indians (£287), Pakistanis (£227), with Bangladeshis trailing in at just £191 (Modood, 1997). A more detailed analysis is given in Table 5.4 (Anderson and Flatley, 1997), which shows the distribution of incomes by quintile. Some 76 per cent of Bangladeshis were in the lowest quintile, compared to 36 per cent of Pakistanis and 30 per cent of Black Caribbean households and just 13 per cent of white households. At the other end of the distribution, 27 per cent of white households were in the top quintile of incomes (and 53 per cent in the top two quintiles), compared to just 2 per cent of Bangladeshi and 7 per cent of Pakistani and Black Caribbean households. The only ethnic groups which approached the white distribution were the Chinese and the Indians.

To some extent these differences reflect the differences in educational attainment between different groups (Heath and McMahon, 1997; Drew *et al.*, 1997; Mortimore *et al.*, 1997). Whereas those of Chinese and Indian origin generally do well at school, and often better than their white compatriots, many Caribbean male students tend to perform relatively poorly.

About a quarter of Black African, Chinese and Other Asian groups had higher educational qualifications in 1991, followed by Indians and Whites with about 15 per cent. The worst qualified were Black Caribbean (10 per cent), Pakistani (8 per cent) and Bangladeshi (5 per cent). The same groups have been shown to have poor school-level qualifications (DfES, 2003).

This may be related to the household structure of some Caribbean families. Scott *et al.* (2001) point to the very large proportion of one-parent Caribbean families compared to other groups. Although the number and proportion of one-parent families has grown dramatically in Britain in the last twenty years (Land and Lewis, 1997) it is particularly high for Caribbean households. As they note:

> The estimates for 1997–99 show that 3 out of 4 children in Great Britain lived with both their natural parents and 1 in 4 lived with either their natural mother or natural father (in the vast majority of cases the mother). However, there were very marked differences between ethnic groups. Asian children aged 0–14 were the most likely to be residing with both parents. Around 9 out of 10 Indians, Pakistani, Bangladeshi, Chinese and Other-Asian (non-mixed) children were living in the same households as their mothers and fathers. In contrast, fewer than half of all Black children and only about one third of Black-Caribbean and Black-Other children shared a home with both natural parents . . . The pattern for White children was intermediate between those for the Black and the Asian groups: three quarters of White children were living with both natural parents.
>
> (Land and Lewis, 1997, p. 13)

This pattern, which appears to replicate family patterns in the Caribbean and amongst Black Americans, is problematic as there is some evidence (Gordon, 1996; Sparkes, 1999) that children in single-parent families tend to perform less well at school and have a greater incidence of behavioural problems. A competing analysis for the poor school performance of young black males attributes it to racism on the part of white teachers who see black pupils as less able, and leads to self-reinforcing behaviour. But this interpretation has been challenged by William Atkinson, the black headmaster of a school in West London, and an appointee to the government task force on raising school standards. He sees the roots of the problem in a growing culture of failure. As he put it in a recent interview:

> Ethnic minority children in our schools are succeeding. But, in some cases, they're underperforming their white counterparts. Part of the problem is the children not belonging to the culture that supports success. They're not coming to school regularly or on time. They've not got the right equipment for lessons. Those who do not succeed buy into a

separate culture – about being smart, the clothes you wear; about being able to do the latest rap. It's all building up against learning. It's not cool to learn. The same would apply to many white working-class children. Interrupting the education of fellow classmates is another form of black-on-black violence but instead of physical violence it's damaging children's chances. Individual students and families must take responsibility and not simply blame the schools. Many schools are working hard with disaffected children to give them the skills and opportunities for success. But too many students are not buying into it. It's too easy to blame the schools.

(Garner and Pyke, 2002)

Ethnic minority unemployment

The level of ethnic minority unemployment in London has been consistently higher than that of the white population. There are, however, substantial differences between ethnic groups. Fieldhouse (1999) notes that 'the highest rates are amongst the city's most recent immigrant groups' (p. 1571). The 1991 census found that Bangladeshis (36 per cent) and Black Africans (30 per cent) had unemployment rates three times the white rate of 10 per cent. Pakistanis (24 per cent), Black Others (24 per cent) and Black Caribbean (19 per cent) had rates of twice or more the white rate. By contrast, those of Indians, Chinese and other Asian ethnic origins had unemployment rates only slightly higher than the white rate. There are also important differences within ethnic groups. Unemployment rates amongst most ethnic groups are highest amongst men and the under-25s and 'the most striking differences are for blacks, who have extremely high male youth unemployment, with levels indicative of a "black youth unemployment crisis" not unlike that experienced in the US' (Fieldhouse, 1999, p. 1571). Among young Black Caribbean, Bangladeshi and Black African males aged 16–24 in 1991, unemployment rates were around 40 per cent. By contrast, the rates for young Black Caribbean women, while still high at 23 per cent, were much lower than the equivalent male rate.

A number of reasons have been advanced for the differential positions of the various ethnic groups in the labour market. One argument, which is widespread in the USA, is that certain ethnic minorities are disadvantaged by virtue of their residential location in relation to the available distribution of jobs. This differential accessibility to jobs is termed 'spatial mismatch'. It is well known that the minority ethnic groups tend to reside in the same areas in which they initially settled, and that the areas are disproportionately concentrated within the economically disadvantaged parts of Britain, particularly the inner cities and industrial towns. The result, Cross and Johnson (2002) suggest, is that 'ethnic minorities are now totally dependent upon the fortunes of inner city economies'. This seems an overstatement

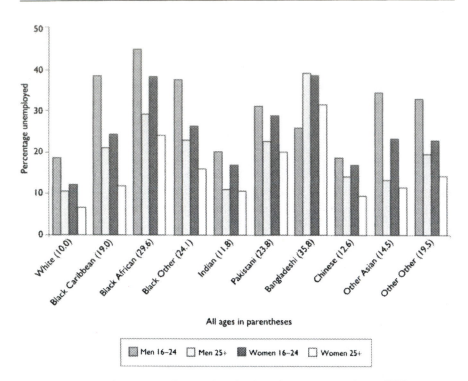

Figure 5.5 Unemployment in Greater London by ethnicity, age and sex, 1991.
Source: Fieldhouse (1999).

given the performance of some ethnic minorities in the labour market, but given the sharp decline of manufacturing industry in Inner London a question emerges whether ethnic minorities have suffered from spatial mismatch of residence and employment (Rogers, 1997).

Propensity to unemployment varies by country of birth, marital status, occupation and housing tenure as well as ethnicity, age and sex, and Fieldhouse (1999) investigated the extent to which these factors could explain some of the differences in unemployment rates using logistic regression modelling. He found, however, that these characteristics 'cannot explain the enormous differences in the risk of unemployment between ethnic groups' (p. 1589). The massive disadvantage in relation to unemployment for all ethnic groups, except for Chinese and other Asians, cannot be attributed to other characteristics even when their social class and housing tenure are taken into account. We are thrown back on other explanations for differences in unemployment, and Fieldhouse examines the extent to which the labour market position of ethnic minority groups may be affected by their residential location if they are concentrated in areas where job opportunities are

fewer and unemployment is higher. Although there is a partial visual correlation between the geography of unemployment in Greater London in 1991
and the pattern of ethnic minorities by ward of residence, he found that the
unemployment pattern of ethnic minorities corresponds to that for the white
population, being highest in the inner areas. He found that it appears to be:

> particularly a feature of areas in which ethnic minorities are heavily
> concentrated in public housing. This, together with the importance of
> the characteristics of ethnic minorities, suggests that it is more to do
> with residential segregation of the most vunerable groups than with
> spatial mismatch of ethnic minorities and jobs.
>
> (Fieldhouse, 1999, p. 1589)

Fieldhouse argued that space may be important through operation of
'concentration' effects, through which a local culture may develop, which
is characterised by a loose sense of attachment to the labour market. This
parallels the findings of W.J. Wilson (1996) in the USA and the points made
by Atkinson (pp. 118–19) regarding education. This interpretation is reinforced by the fact that not all ethnic minorities are doing badly in school
or the labour market. On the contrary, some are doing very well, which
suggests that the key issue is one of differential performance, as Modood
(1997) argues.

Ethnic minorities in the London housing market

Poor immigrants almost inevitably start off their housing careers in the
worst housing because they lack the education, the skills, the contacts and
the language necessary to gain access to good jobs and reasonable incomes.
This was seen in nineteenth-century New York, where Jewish migrants were
concentrated on the Lower East Side in poor-quality tenement housing,
and in nineteenth-century London where the recent Jewish immigrants
lived in overcrowded conditions in the East End, just as the Bangladeshis
do today. There was a persistent tendency to blame immigrants for their
poor and overcrowded housing conditions, ignoring the fact that because
they are poor they can afford little else. Even Engels (1969) was not exempt
from this, and his comments on the culture and housing conditions of the
Irish immigrants in nineteenth-century British cities are so overtly racist
that they do not bear quotation (interested readers should see Engels, 1969,
pp. 122–5).

The disadvantaged position of ethnic minorities in London's labour
market has been reflected in their position within the housing market. When
the first waves of New Commonwealth migrants arrived in the 1960s they
generally lacked the money to buy and were ineligible for council housing
because they lacked the minimum residence requirements then in force.

When they managed to get onto council waiting lists, they would generally have a long wait as points were allocated for length of time on the waiting list, amongst other things. As a result, New Commonwealth immigrants in London and the other big cities almost invariably found themselves in poor-quality, overcrowded private-rented accommodation where they faced discrimination and sporadic harassment (Sarre *et al.*, 1989; Milner Holland, 1965). Ramdin (1987) notes that 'West Indians were pushed into a decaying environment which became their permanent homes. In effect, prejudice and discrimination limited their opportunities for better accommodation' (p. 194).

After 1970 the housing careers of New Commonwealth migrants began to diverge. While some West Indian migrants managed to buy houses in the cheaper areas of London, the majority began to make headway in getting into the council-rented sector, particularly after 1974 when the discriminatory barriers to entry began to be removed as a result of the Cullingworth Report. But Indian and Pakistani migrants who had managed to accumulate some capital, often by borrowing from families, began to buy owner-occupied housing. These differences intensified during the 1980s, and by 1991 the housing market positions of the various groups of migrants, and their descendants, were very different. Whereas those of West Indian ethnic origin became concentrated in the council sector, those of Indian and Pakistani origin have achieved a very strong position in home ownership.

The result of their differences in labour market and housing careers is that different ethnic groups now exhibit very different patterns of housing tenure. At one extreme, two-thirds of the Bangladeshi population were concentrated in the social rented sector with only 27 per cent in home ownership. They are closely followed by Black Africans with 59 per cent in social rented housing and 25 per cent home ownership. The Black Caribbean group have an intermediate position with 50 per cent in social housing but 45 per cent home ownership. At the other end, 80 per cent of the Indian population (and 68 per cent of Pakistanis) are in home ownership, with just 13 per cent and 21 per cent respectively in social renting. The white group, because of their numerical importance, closely conform to the average for all groups with 59 per cent in home ownership and 28 per cent in social renting.

To some extent, the concentration of Bangladeshis and Black Africans in the social rented sector is a reflection of their more recent arrival, and the fact that they have not yet become well established in the labour market. This may be partly rectified in time, but the discrepancy in housing tenure patterns between the Caribbean and the Indian and Pakistani groups suggests that the Caribbean group has failed to make progress into home ownership achieved by the other groups. More problematic is the fact that those of West Indian and Bangladeshi ethnic origin tend to be concentrated in the worse sections of the council sector, either in high-rise, high-density estates, or in older, poor-quality property. Peach and Byron (1993) have shown that those of West Indian origin are disproportionately concentrated both in

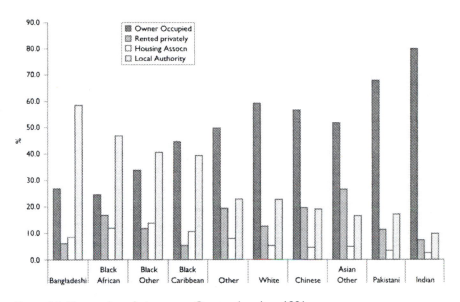

Figure 5.6 Tenure by ethnic group, Greater London, 1991.
Source: Census of Population, 1991.

high-rise blocks and on the higher floors of those blocks. Parker and Dugmore (1978) looked at the allocation of GLC housing, and Phillips and Karn (1992), looking at allocation to minorities in Tower Hamlets, also found that ethnic minorities tended to be allocated poorer-quality housing or worse estates than the indigenous population (Henderson and Karn, 1984).

The reasons for this are complex, but involve discrimination in terms of allocation of properties and the interaction of housing needs, and in the operation of the allocation system. A disproportionate number of West Indian applicants for council housing in Inner London tend to be single mothers. They may be homeless or in acute housing need, and they need housing immediately. But, because better-quality housing is less likely to be vacated than poor-quality housing in problem estates, the housing stock available for re-letting at any point in time is more likely to contain a disproportionate concentration of poor-quality stock which has been vacated. Consequently, applicants in urgent need of housing who do not have the luxury of being able to pick and choose among vacancies are faced with taking the first vacancy which becomes available and is offered to them. These are very likely to be the less-desirable properties. Thus those in greatest housing need tend to get concentrated in the poorest and less desirable properties simply because they are likely to be the ones available (Figure 5.7).

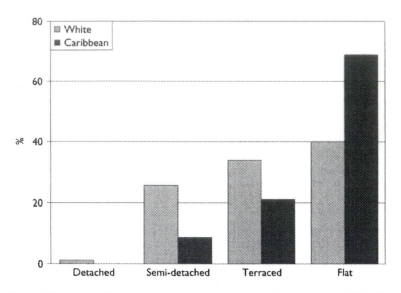

Figure 5.7 Local authority housing type by ethnicity, Great Britain, 1986–9.
Source: Peach and Byron (1993).

The distribution of ethnic minority groups within London

We have looked at the growing proportion of ethnic minorities in London and in Inner London. They do, however, have a distinct spatial concentration in London, with high figures in some boroughs in 2001, particularly in Brent (35 per cent), Newham (51 per cent), Tower Hamlets (36 per cent), Hackney (42 per cent), Ealing (41 per cent), Lambeth (38 per cent) and Harrow (34 per cent). The geographical distribution in 1991 is shown in Figure 5.8. The ethnic minority population of Inner London grew by 48 per cent and that of Outer London by 59 per cent. This growth raises the key issue of whether ethnic minorities in London are becoming more concentrated into a small number of disadvantaged areas. The answer to this is a clear 'no' on two fronts. First, the distributions of different ethnic groups are very different, with those of Asian origin concentrating more in Outer London, and, second, as Peach (1996) has shown, ethnic groups are tending to disperse outwards from their areas of initial concentration as a result of the housing allocation process (Peach and Shah, 1980). However, there may be a concentration of minority groups in certain local authority estates. What seems to be happening is that as the ethnic minority population of London grows it is simultaneously expanding outward geographically. This means that an increasing number of wards have a higher ethnic minority population, but this is not ghettoisation in the traditional sense of the word.

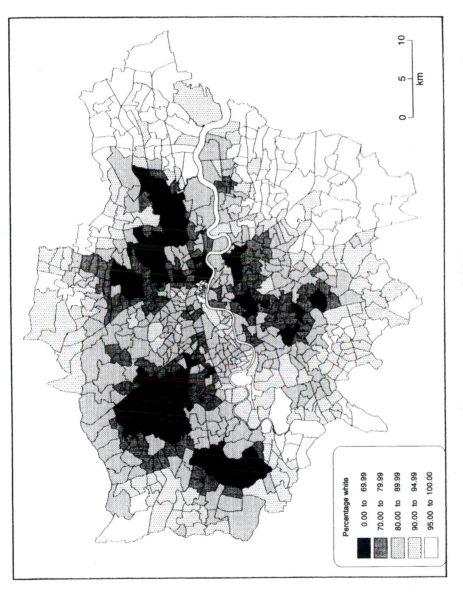

Percentage white

■	0.00 to 69.99
	70.00 to 79.99
	80.00 to 89.99
	90.00 to 94.99
	95.00 to 100.00

Figure 5.8 Ethnic composition in Greater London, by ward, 1991.
Source: Fieldhouse (1999).

As Deakin pointed out thirty years ago:

> The geographical distribution of New Commonwealth immigrants in 1971 does not follow the 'classic' immigrant model of concentration in the inner city. Instead, the immigrants are relatively widely distributed at borough level throughout a range of areas, a pattern partly determined by the availability of cheaper housing to rent or buy, and partly by employment opportunities . . . Different groups within the population also have quite marked differences in distribution. It is therefore probably more helpful to think in terms of series of 'ethnic clusters' scattered over a series of London boroughs.
>
> (Deakin, 1974)

More recently, Peach, comparing pluralist and assimilationist models of ethnic settlement in London, suggests that:

> both models hypothesise that ethnic minority populations would manifest significant degrees of ethnic segregation in working class areas in their early days of settlement, but that as economic development takes place, assimilating groups would diffuse from their areas of initial settlement, while pluralist groups would maintain enclaves even if the enclaves themselves relocated.
>
> (Peach, 1997, p. 120)

Peach argues that the social geography of London manifests both models, with the Caribbean population moving along the assimilationist trajectory and the South Asian population following a more pluralistic path. He points to the fact that the segregation rates for the Caribbean-born population have shown a consistent decline from 1961 to 1991, whereas rates for the Asian population showed slight increases or stability. He notes that the highest concentration of ethnic minorities in London wards are formed by Indians and Bangladeshis, with peak concentrations twice as high as Caribbean ones. He points out that the Caribbeans are the only group for which there is unambiguous evidence of structural assimilation. No fewer than 20 per cent of Caribbean couple households in London consisted of mixed black and white couples in 1991.

Conclusions

London performs the same role as New York and Los Angeles as a major gateway city for overseas migrants, and the rate of migration has increased in recent years. Forty years ago London was still predominantly a white mono-ethnic city, although it possessed significant groups of Irish, Jewish and Cypriot residents. Today, there can be little dispute that London is one of the most multi-ethnic cities in Europe, although it has not yet reached the levels of ethnic diversity found in New York or Los Angeles. The size of the ethnic minority groups continues to expand, and refugees and asylum seekers are adding to the diversity. The differential age structure and birth rates of some ethnic groups will serve to increase the proportion of ethnic minority population over the next decade. As in other major cities, the key issue today is not the systemic failure of all ethnic minorities to succeed economically and socially, and overcome discrimination and disadvantage, but the differential success of certain groups. As in the USA, the Indians and the Chinese are moving steadily into management and the professions, like Jewish migrants before them. But, in the process, many of those of Black Caribbean origin, along with many members of the less-skilled white working class, are being left behind. The issue is to ensure that these groups can be effectively drawn back into education and the labour market, lest they become permanently left behind, isolated and alienated from mainstream society.

Publication of the preliminary results of the 2001 census has highlighted the dramatic transformation which has taken place in the ethnic structure of London in recent decades. The non-white ethnic minority population of London in 2001 was just over 2 million or 29 per cent: an increase of 722,000 or 53 per cent from 1991. In Inner London ethnic minorities grew from a quarter of the population in 1991 to just over a third (34 per cent) in 2001, and in Outer London from 17 per cent to 25 per cent. In Newham (61 per cent) and Brent (55 per cent), ethnic 'minorities' have become the majority population, and Tower Hamlets (49 per cent) is close. All boroughs have seen the growth of their ethnic minority population, ranging from 13 per cent in Wandsworth to over 80 per cent in some boroughs. Some of the biggest increases have been seen in suburban rather than inner boroughs. Simultaneously, the white population of London fell by 4 per cent overall. The biggest decreases were Harrow (−18 per cent) and Newham (−21 per cent), but Croydon, Waltham Forest and Redbridge all had decreases of over 10 per cent. What this suggests is a process of outwards ethnic minority expansion and succession in London, with white residents moving outside London or further afield. Suburban Harrow, for example, lost around 27,000 white residents in the 1990s, while non-whites grew by 32,000. Croydon lost 26,000 whites and gained 43,000 non-whites and Redbridge and Waltham Forest showed similar trends. If the ethnic minority population of London was to grow in the current decade at the same rate it did in the 1990s (+53 per cent), and the white population shrinks by 4 per cent, ethnic minorities would total about 3.16 million by 2011, compared to a white population of 4.9 million: just under 40 per cent and Inner London would have an ethnic minority population of 1.4 million – about 44 per cent – compared to 1.77 million whites. This suggests changes which will make London increasingly like New York and Los Angeles in terms of ethnic diversity. The predominantly white, mono-ethnic London of the 1960s is now a thing of the past.

Chapter 6

The transformation of the housing market

The turn of the century seems likely to see the whole of inner London parcelled out either into council blocks . . . or into highly expensive and inevitably exclusive owner occupied property. The gain in overall appearance and repair may be considerable, the loss in flexibility and in the possibility of living space for those who are neither well-off nor clever, will be immeasurable.

(Tindall, 1973)

Between the extremes of the yuppie couple making their fortune and the single parent living in a decaying council estate, Hackney encapsulates what has been happening to London. Our capital city is now inhabited by two different populations: there is a manichean division between winners and losers, the upwardly mobile and the no-hopers, mortgage payers and rent payers. The division is growing sharper, and it is taking place as a result of *policy*.

(Bugler, 1985, p. 7)

any district in or near London, however dingy, or unfashionable, is likely to become expensive, and London may quite soon be a city which illustrates the principle of the survival of the fittest: the financially fittest, who can still afford to work and live there.

(Glass, [1963] 1989, pp. 140–1)

Introduction

Gillian Tindall's comments on the polarisation of the Inner London housing market published in *New Society* in 1973 have proved remarkably prescient. At that time private renting was still important but declining and it was possible to see at first hand the rapid expansion of gentrification and of new council estates, and the emergence of the housing tenure segregation she described. In the thirty years since she wrote it has become increasingly clear that, like Ruth Glass, she was fundamentally correct in her predictions.

Like other global cities, London has become polarised into two distinct housing markets: on the one hand, the increasingly expensive market where prices are set at the top end by the demand from highly paid workers in finance, law and related areas, as well as the international wealthy, and on the other, the council and social rented sector which houses most of those who cannot gain access to home ownership or high-cost private renting. The social divide between the tenures has grown rapidly since the 1960s, leading to the increasing social residualisation of council housing and an increased social polarisation between council tenants and homeowners.

The roots of the transformation of London's housing market lie in four interrelated processes. The first is the growth of the professional and managerial middle classes and the increased demand for attractive housing they generate by virtue of their greater purchasing power. The second is the sharp decline in the private rented sector, which was eroded both as a result of local authority clearance and redevelopment and the sale or conversion of better-quality property for owner-occupation. The third and fourth processes are the converse of the second: first, the dramatic growth of council housing in the 1960s and 1970s, and second the rapid growth of home ownership in Inner London from the 1970s onwards. These processes and the associated movement of different groups between tenures have reshaped the physical form, the tenure and social structure of the housing market and accessibility to housing. The chapter looks in detail at the tenurial and social changes which have taken place, but first it is necessary briefly to discuss the nature of the housing market in a market economy.

The allocation of housing in market economies

In market economies private housing is a commodity produced, sold and exchanged for profit. Outside the social or public housing sector where housing may be allocated on the basis of some criterion of need or at a below-market price, housing is competed for through the market, where it is allocated on the basis of price and ability to pay. It follows that, for the majority of buyers who do not possess a private income or inherited wealth, access to private housing is significantly constrained, if not totally determined by earned income and hence by the position of individuals within the paid labour market. As housing comprises a very important element within most people's lives, it follows that most people will want to buy or rent the most attractive housing possible within their means. But, as the supply of attractive housing in desirable areas is limited, the most attractive housing will always command the highest price. There is therefore a hierarchy of desirability within cities, with the most attractive housing in the best areas commanding the highest price, and so on down the scale to the least attractive housing in the least attractive areas (Hamnett and Randolph, 1988b).

This is not to say that everyone wants to live in Kensington, but that enough people with the money do to keep prices high.

In his book *Social Justice and the City*, David Harvey (1973) used the analogy of a theatre where the seats are allocated by ability to pay. The first person into the theatre with the highest income is able to choose the best seat, and so on down the line, until the last person into the theatre finds themselves at the back with a restricted view and an uncomfortable seat. People who lack even the minimum price of admission find themselves on the street. This should be no surprise to anyone with experience of the private housing market. Those with low skills, low incomes or the long-term unemployed are unlikely to be able to gain access to the private housing market except at the lowest levels. The operation of the price mechanism serves to exclude those who do not have the financial resources to compete. Conversely, those with the highest incomes and/or the greatest wealth are generally found in the most attractive and expensive housing in the best areas. As Paul Harrison (1983) put it: 'British cities are segregated by class . . . No laws of apartheid are needed to enforce this separation: it occurs as naturally as oil divides itself from water, by the play of unequal incomes and savings in the housing market' (p. 180).

Although it is possible to quibble with Harrison's use of the term 'naturally', his basic proposition is undoubtedly correct in market economies where labour-market position and earnings are decisive in influencing position in the housing market. The key point is that where people live is not simply a product of free, unconstrained choice. Rather, it is a product of the interaction of the changing structure of housing supply, knowledge of the housing market, residential preferences and ability to pay. As shown later in the chapter, there is a strong link between class, housing tenure and the price of housing.

Position in the labour market is not, of course, the sole determinant of housing market position. Given the necessity for most people to raise a mortgage to buy a house, the lending policies of building societies and banks are important and, until about the late 1970s, they discriminated against certain types of household, notably single women and ethnic minorities, and particular areas and types of property where the nature of the housing stock was considered a bad risk. Inner-city terraced housing was generally not well regarded. Estate agents also attempted to guide certain groups to certain areas and some sellers refused to sell to ethnic minorities. Until introduction of legislation which outlawed it, private landlords frequently discriminated against minorities, and within the local authority sector discrimination has historically been important in either denying access to minority groups or allocating them to marginal properties in marginal estates in marginal areas (Henderson and Karn, 1984; Phillips, 1987; Parker and Dugmore, 1978; Peach and Byron, 1993; Sarre *et al.*, 1989).

Housing and labour-market position

Given this relationship between the labour and housing market, the argument to be made here is very straightforward. First, that position in the London housing market is strongly related to labour-market position and income, and second that the changes in London's industrial, occupational and earnings structure have had a dramatic impact on the nature of the housing market. Put simply, the sharp increases in the size of the professional and managerial occupational groups over the last thirty years, combined with the very large increases in earnings and incomes that this group have had, have, along with the declining numerical importance of manual workers and the growth of economic inactivity, significantly changed the structure of housing market demand in London. The second point follows from this: given the changing structure of housing supply, there has been a growing social polarisation between housing tenures.

Merrett (1988) argues that owner-occupied housing in London can be characterised in Ricardo's terms as one of those scarce commodities 'the value of which is determined by their scarcity alone'. Such commodities have the attribute, along with rare art works and the like, that the supply cannot be rapidly increased, or can only be increased with considerable difficulty and cost. Merrett argues that the supply of housing in London is able to increase only slowly with restrictions on new building and conversion and that as the price is largely determined by demand, increase the demand (or the ability to pay) and the price will rise accordingly. This has been intensified by London's role as a centre of global finance and the growing number of small households. The resultant rise in owner-occupied house prices directly reduces the comparative yield of private-rented property held by absentee landlords, and their supply responses 'intensify the effective demand for owner occupation from households which cannot gain access to the rental sector' (Merrett, 1988, p. 17). People living outside London often marvel at the level of house prices and ask who can afford to pay them. Like Paris, Tokyo and New York, the answer lies in the occupational and earnings structure of the city. We have already shown the major transformation in the industrial, occupational and earnings structure of London and, not surprisingly, the growth of affluent middle-class demand has pushed up house prices substantially across the London area, particularly in those parts of Inner London which are most attractive to the growing middle class who work in central London and want easy access both to work and to entertainment facilities.

The next section examines the historical structure of housing in London until the late 1960s, and then we examine the changes which have taken place since the 1960s – first the rise of the council sector, then the rise of owner-occupation and the growing polarisation between the two tenures. Finally, the chapter turns to look at the scale of house-price inflation and London and the problems it poses for affordability.

The historical legacy

The structure of Inner London's housing market has changed dramatically over the last forty years. Whereas, in 1961, Inner London was still overwhelmingly dominated by private renting, with a small council sector and an even smaller owner-occupied sector, the Inner London housing market has been almost completely reshaped. The private rented sector is now a tiny shadow of its former self, owner-occupation is now the largest tenure, and the council rented sector, which expanded enormously during the 1960s and 1970s to a peak in 1981, has contracted as a result of right-to-buy sales and large-scale estate transfers to housing associations and the private sector. Outer London, by contrast, was primarily built for home ownership in the inter-war years, and has become increasingly dominated by home ownership ever since. Consequently, a growing tenure and social divide opened up between Inner and Outer London during the 1960s and 1970s. This tenure divide has subsequently narrowed as ownership has grown in Inner London.

To understand the nature and impact of these changes, and the differences between Inner and Outer London, it is useful briefly to go back to the early twentieth century before the rise of mass suburbanisation, when London stretched 7–8 miles out from the centre, compared to the 15 miles or so today. The London of that period consisted almost entirely of private rented terraced houses, with very little in the way of owner-occupation and virtually no council housing: both of which were largely products of the 1920s onwards (Jennings, 1971; Jackson, 1974). The late Victorian housing boom had already led to the expansion of suburban London, aided by expansion of the railways, the tube system and the electric tram (Prince, 1964). As Jenkins put it:

> The spread of development up the main lines was remorseless. The Great Northern Railway into King's Cross drew the pressure of population off the mid-Victorian suburbs of Hackney, Islington and Holloway out into the salubrious estates of Palmers Green, Wood Green, Bowes Park and Hornsey. The London and North-Western, with its magnificent terminus at Euston, at first avoided suburban traffic . . . But its Broad Street extension and the famous 'North London Line' from Richmond, opened up large tracts of suburbia in the 1860s – Highbury, Kentish Town, Gospel Oak, West Hampstead, Kilburn, Kensal Green, Willesden, Acton, Chiswick and Kew . . . South of the River the same phenomenon was maintained.
>
> (Jenkins, 1975, p. 131)

The inter-war years saw the continuing rapid outward expansion of London, primarily in the form of speculatively built owner-occupied suburban houses around a privately rented nineteenth-century core. The scale of this expansion was remarkable (Johnson, 1964; Jackson, 1974; Jenkins, 1975) and the area of Greater London doubled between the wars. This was the era of John Betjeman's 'Metroland' and it saw the development of a 'semi-detached London' unlike the terraced houses of Inner London.

The transformation of the 1960s and 1970s: the rise of council housing

From the 1960s onwards this distinction between a primarily privately rented Inner London and owner-occupied suburbia was to change dramatically as a result of both clearance and local authority redevelopment of poor-quality privately rented stock and the growth of ownership in Inner London. At the end of the war, a large proportion of the existing housing stock in inner east London around the docks had been damaged or destroyed. In addition, much of the private rented housing was overcrowded, poorly maintained and poorly provided with basic facilities, with high levels of households sharing accommodation. As a result, Labour councils in Inner London developed the inter-war policy of large-scale clearance and redevelopment of the worst areas of private renting in the form of high-density council estates (London County Council, 1937). Their aim was to improve housing and living conditions for the working classes (Dunleavy, 1981; Merrett, 1979; Young and Kramer, 1978).

It is difficult today to appreciate the magnitude of the poor housing conditions in Inner London in the 1960s which led to a Royal Commission of Inquiry and publication of a report on Housing in London in 1965 (Milner Holland, 1965). In the early post-war period many rented houses in Inner London were split into rooms or unselfcontained flats and, as a consequence, many households had to share basic facilities and slum landlordism was a major problem (Donnison, 1967). Almost two-thirds of households (64 per cent) still lived in shared dwellings in Inner London in 1951, but this fell to 44 per cent in 1961. In 1961, 7 per cent of households still lacked exclusive use of both a cooker and a sink, 18.6 per cent shared a bath with other households, and 30.5 per cent were entirely without use of a fixed bath. In addition 30 per cent of households shared a toilet with other households and overcrowding was a major problem. In 1951 12 per cent of households lived at a density of over 1.5 persons per room, and in 1961 the figure was still 11.5 per cent. These conditions were primarily concentrated in the privately rented sector and there were major problems with landlords who made life a nightmare for some tenants. One of the most notorious of these landlords, Perec Rachman, gave his name to the term 'Rachmanism' which is now a synonym for rapacious and exploitative landlordism.

It is not surprising that one of the major goals of the predominately Labour-controlled local authorities in Inner London was to clear as much of the poor-quality private rented sector as possible and redevelop it in the form of new local authority estates. In this way, they believed that they could improve housing conditions for the mass of the population, and housing conditions improved dramatically during the 1960s and 1970s.

Initially, the new council estates did just this. By 1991 no less than 97 per cent of households in Inner London had exclusive use of a bath or shower, and under 0.5 per cent had shared use or lacked an inside WC. In terms of internal facilities council housing was a massive improvement on conditions in the private rented sector and it broke the link between poverty and poor housing. The subsequent problems of the council sector have lain not in their internal facilities but in the overall design and construction of buildings and estates, particularly in the shift to high-density and high-rise buildings, many of which have proved physically and socially problematic.

There is no doubt that many Inner London Labour councils thought at the time that they were building a new world of high-quality housing for the local population. The result, however, is a built environment which because of the adoption of prefabricated concrete panel construction, and inherent design faults such as the notion of 'streets in the sky' and the reliance on lifts and long walkways, with minimal private space, created spaces which quickly deteriorated. Coleman (1985) strongly criticised the environmental and social results in her polemical book *Utopia on Trial*. Although her analytical approach has been strongly criticised for its physical determinism and its failure to examine the social composition of the estates, she nonetheless put her finger on a fundamental flaw in the design of many of these estates. A number of them have subsequently been subject to major modifications, including removal of higher floors and the aerial walkways which were frequently an invitation to crime, and some have been demolished. Many estates still remain, however, and the large areas of unsupervised communal space, crime and vandalism, have resulted in a reduction in quality of life for many residents (Harrison, 1983; Power, 1993, 1996).

Unfortunately, although council housing initially proved highly desirable and a big improvement on the quality of the private housing it replaced, the growth of home ownership, combined with the out-migration of many members of the skilled working class from Inner London (Deakin and Ungerson, 1977) and the sale of the better parts of the council stock, has meant that the social composition of council estates has changed considerably. Many of the original skilled tenants have moved into home ownership to be replaced by the less skilled, the unemployed, the poor, single parents, the homeless and some ethnic minority groups (Henderson and Karn, 1984; Hamnett and Randolph, 1987; Forrest and Murie, 1990). At the same time, the physical quality of many estates has deteriorated and they have become a locus of crime and violence (Gray, 2001). In parts of Inner London, as in other cities, council estates have become the embodiment of the urban 'other' (Watt, 2001), a locus of social exclusion and social problems. The problems were highlighted by the case of Damilola Taylor, a young Nigerian boy aged 12, who was stabbed to death in a council estate in North Peckham on his way home from school in 2000 (Sengupta, 2000; Cobain, 2000).

Whereas the rise, and partial decline, of the council sector is a result of government policy and the determination of local councils to clear and redevelop large areas of poor-quality private rented housing, the growth of home ownership owes a great deal to the dramatic transformation of the industrial, occupational and income structure of London. The rapid growth of the new class of professionals, managers, technical specialists and other highly skilled service workers has led to a dramatic increase in the demand for home ownership. The resultant changes in the economics of private landlordism made it more profitable to sell for home ownership than to continue renting (Hamnett and Randolph, 1987), and the gradual changes in building society lending policy in the 1960s and early 1970s led to more mortgage lending for inner-city home ownership (Williams, 1976, 1978).

The scale and geography of the housing market transformation

In 1961 almost two-thirds (64 per cent) of all households in Inner London were private rented tenants, and in the central boroughs of Camden, Westminster and Kensington and Chelsea the proportion was four-fifths. Owner-occupation was small scale at around 17 per cent and council housing accounted for 19 per cent, much of it in the form of older London County Council housing estates built between the wars. Inner London at this date was still overwhelmingly privately rented. In Outer London, by contrast, private renting accounted for just 30 per cent of all households: less than half the proportion in Inner London. Conversely, home ownership accounted for 53 per cent of all households in Outer London: more than three times the proportion in Inner London. Council housing in Outer London was, at 17 per cent, only marginally below the level in Inner London. There was therefore essentially a privately rented core with a suburban ring of ownership and limited council housing in both Inner and Outer London (Figures 6.1, 6.2).

The decades from 1961 to 1991 saw a dramatic transformation in the tenure structure of Inner London. By 1981 the proportion of households in private renting had fallen by half from almost two-thirds to just under a third (32 per cent), while the council rented sector had grown to become the single largest tenure at 42 per cent as a result of large-scale clearance and redevelopment by the local Labour councils (Figure 6.1). In a number of boroughs such as Hackney (57 per cent), Islington (56 per cent), Tower Hamlets (82 per cent) and Southwark (65 per cent), council renting accounted for over 50 per cent of households in 1981 (Figure 6.3). Home ownership grew much more slowly to reach 27 per cent by 1981. The 1960s and 1970s were decades of rapid council housing expansion in Inner London

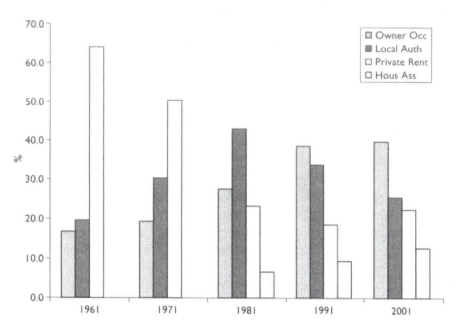

Figure 6.1 Household tenure, Inner London, 1961–2001.
Sources: Censuses of Population, 1961, 1971, 1981, 1991 and 2001.

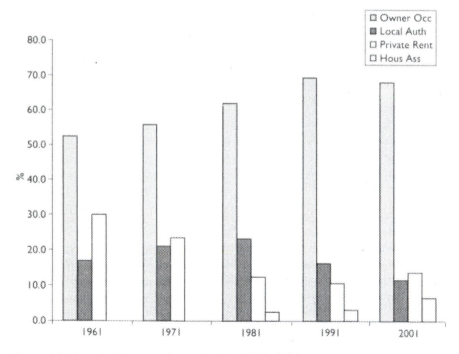

Figure 6.2 Household tenure, Outer London, 1961–2001.
Sources: Censuses of Population, 1961, 1971, 1981, 1991 and 2001.

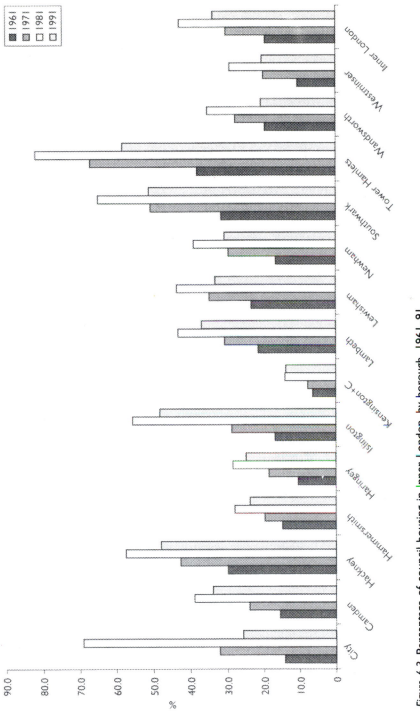

Figure 6.3 Percentage of council housing in Inner London, by borough, 1961–91.
Sources: Censuses of Population, 1961, 1971, 1981 and 1991.

with a growth of 75 per cent over the two decades. In some boroughs the growth was even more dramatic: 96 per cent in Camden, 100 per cent in Westminster, 104 per cent in Newham, 129 per cent in Islington and 133 per cent in Haringey (Figure 6.4).

This was the high watermark of the sector. As a result of the introduction of right-to-buy policies, estate sales, and a dramatic contraction of new council building under Margaret Thatcher's Conservative government from 1979, the council sector began to shrink during the 1980s and by 1991 it had fallen by just over 10 per cent in absolute terms to 34 per cent. The rate of decline during the 1990s was considerably greater as a result of intensified right-to-buy policies and large-scale estate transfers to other social landlords, primarily housing associations. The 2001 census showed that council housing accounts for only 25 per cent or less of households in Inner London, while home ownership had grown slightly to 40 per cent and private renting had recovered to its 1981 level.

In the space of forty years Inner London has been transformed from a city dominated by private renting, to one where council renting became the single largest tenure in 1981, to one where home ownership is the largest tenure. The growth of ownership in Inner London has speeded up each decade from 1961. In the 1960s home ownership grew by only 8 per cent, in the 1970s it grew by 21 per cent, and in the 1980s it grew by a remarkable 60 per cent from 263,000 to 423,000 households. Although the home ownership market slump of the first half of the 1990s slowed this growth down dramatically, it grew again in the second half of the decade to 484,000 households by 2001. The increases in some boroughs have been even more dramatic. In Hackney and Islington, ownership grew 80 per cent during the 1980s, in Southwark by 100 per cent and in Tower Hamlets by an astonishing 500 per cent, though from a very low base (Figure 6.4). This latter figure reflects the dramatic transformation of Docklands and the construction and conversion of large numbers of new units for sale. Looked at overall, in the space of thirty years, Inner London has been successively dominated by three different tenures: private renting, council housing and home ownership. The surprise of the 1990s has been the recovery of private renting as a result of government legislation and buy-to-let.

The advent of suburbanisation in the inter-war period coincided with the growth of owner-occupation. As a result, the majority of houses and flats built in suburban Outer London were built for sale, and even in 1961 a majority (53 per cent) of households in Outer London were in owner-occupation. The council sector was relatively small because households in poor housing conditions were concentrated in the private rented sector in Inner London and the Conservative-controlled Outer London councils did not want to provide any more council housing than was strictly necessary for their own housing needs, particularly if it would attract Labour-voting, inner-city working-class tenants (Young and Kramer, 1978; Saunders, 1979).

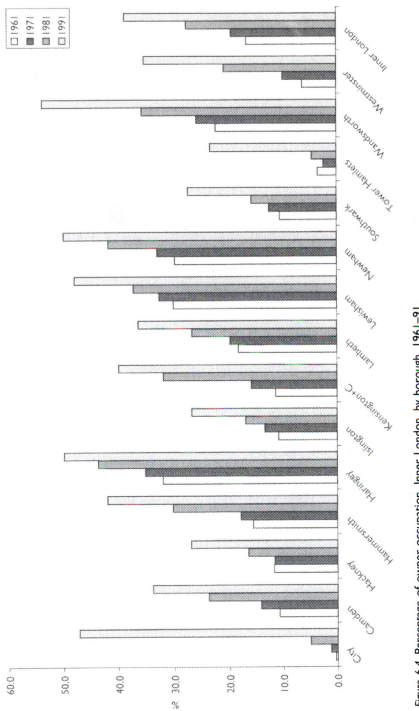

Figure 6.4 Percentage of owner-occupation, Inner London, by borough, 1961–91.
Sources: Censuses of Population, 1961, 1971, 1981 and 1991.

The tenure structure of Outer London has therefore been characterised by a steadily increasing proportion of owners: from 53 per cent in 1961 to 69 per cent in 1991. Despite the continuation of council house sales under the 'right-to-buy' legislation during the 1990s and large-scale construction of suburban blocks of flats, home ownership in Outer London in 2001 fell slightly to 68 per cent as a result of private renting.

From private renting to home ownership: forms of transformation

The transformation from private renting to owner-occupation in Inner London has taken a number of different forms: first, the sale for owner-occupation of individual privately rented houses, sometimes to sitting tenants; second, the break-up and sale for owner-occupation of purpose-built blocks of privately rented flats; third, the conversion of houses, many of which were in multiple occupation, into flats for sale; fourth, the conversion of previously industrial or commercial buildings into owner-occupied flats, fifth, the sale of local authority dwellings under right-to-buy and other legislation; and, finally, new building for owner-occupation. It is not too great an exaggeration to suggest that in many areas of Inner London anything that could be converted into luxury private flats has been or is being converted: schools, hospitals, factories, warehouses, nursing homes, office blocks, the former GLC headquarters and even an old soup kitchen for the poor in the East End which is now luxury apartments for City workers. This point is developed in Chapter 9.

The sale of individual houses

The sale of individual privately rented houses for owner-occupation has been a large-scale phenomenon in London which started in the 1950s and picked up momentum during the 1970s, both as a result of continuing rent control which held down returns for landlords, and the sharp rise in house prices which offered landlords a profitable way to exit renting. The role of increasing house prices cannot be understated and it owes a great deal to the growing number of new middle-class households in London.

The sale of individual houses is traditionally linked with the rise of gentrification; that is, the replacement of existing working-class tenants by the in-movement of middle-class owners, though in a number of cases the properties were sold to sitting tenants. But this process was not able to start until building societies began to relax their lending criteria and offer mortgages on inner-city properties. As Williams (1976, 1978) has shown, this relaxation in policy opened the door to gentrification in areas of Inner London such as Islington. The great majority of former privately rented houses in London have now been sold for individual home ownership or converted into flats (Holmans, 1987). The result has been a substantial reduction in residential densities as multiple occupancy gave way to single-family occupancy.

Flat 'break-ups'

The 1870s saw the development of blocks of rented mansion flats in central and Inner London. In the years up to the First World War the number of blocks increased dramatically, and there was a second wave of building during the 1930s. As a result, these privately rented blocks of flats accounted for 28 per cent of all privately rented dwellings in Greater London in 1966 immediately prior to break-up and sale. In Inner London, the figure was 33 per cent, and in the three central London boroughs it was a remarkable 48 per cent (Hamnett and Randolph, 1984, 1986a, 1988a). Many of these blocks were owned by large institutional landlords, including insurance companies, and by private property companies, and many, particularly in central London, were large and very luxurious bastions of upper-middle-class renting. The geographical distribution of purpose-built private flats and flat conversions compared to the distribution of single-family houses in London in 1981 is shown in Figure 6.5. There is a clear 'inner–outer' division.

The 'break-up' and sale for owner-occupation of blocks of previously privately rented flats got under way in Inner London in the 1960s as a result of three principal factors. The first was the continuing legacy of rent control which depressed landlord yields. Second, the continuing scandals surrounding the cowboy-end of private landlordism reduced the attraction for large institutional landlords to remain in rented property. Third, the rapid inflation in house prices, particularly in the early 1970s, offered landlords the prospect of substantial capital gains from sales. Hamnett and Randolph (1988a) have shown that break-ups made a major contribution to the decline of private renting and the growth of home ownership in central London during the 1970s and 1980s. Most of the formerly rented blocks have now been largely converted to owner-occupation and little potential now remains. The process of transformation, however, was marked by considerable conflict between landlords, agents and tenants, as many institutional landlords sold their rented blocks to speculative owners, many of whom were registered overseas and whose objective was to realise as much profit from the deal as possible. Rents, service and maintenance charges were put up dramatically, and tenants were often offered money to quit so the landlord could sell vacant units.

The root of the problem lay in the nature of English property law, in particular the distinction between freehold and leasehold ownership. Freehold owners own both the property and the ground on which it is built. Leasehold owners purchase a long lease (generally of 99 or 125 years) on the property or part of the property – for example, an individual flat – and at

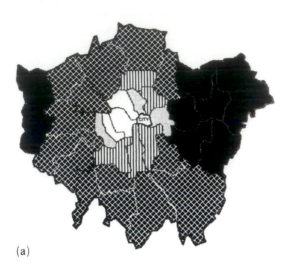

Percentage
☐ < 50
▨ 50 – 69
▥ 70 – 79
▧ 80 – 89
■ ≥ 90

(a)

Percentage
☐ 0 – 5
▨ 6 – 9
▥ 10 – 19
▧ 20 – 29
■ ≥ 30

(b)

Percentage
☐ 0 – 5
▨ 6 – 9
▥ 10 – 19
▧ 20 – 29
■ ≥ 30

(c)

the expiry of the lease the property reverts to the freeholder who is free to sell a long lease all over again. In addition, leaseholders have to pay annual 'ground rent' for their property, plus annual service and maintenance charges. But it is up to the freeholder to make arrangements for the proper servicing, maintenance and insurance of the building, and to bill leaseholders or rented tenants accordingly. Needless to say, the potential for under-maintenance, poor services, over-charging and insurance fraud were virtually boundless, and the new speculative landlords, many of which took the form of nominee companies based in overseas tax havens, who had bought rented blocks to 'break-up' and sell for owner-occupation, saw these charges as simply another way to make money. They frequently had no contact with the long lease-holders or tenants except through intermediaries, and they had no address in the UK on which disgruntled tenants could serve legal notices.

Eventually, the problems became so intense that the Conservative govern-ment set up an independent committee of inquiry into the issue, resulting in the Landlord and Tenant Act, 1987 which tightened up controls on the process and received all-party support in Parliament (DoE, 1985). In many ways, however, the legislation did not go far enough and was a classic case of attempting to lock the stable door after the horse of private profit had bolted. A combination of the private property lobby and government timidity ensured that the legislative powers were weaker than they should have been.

Flat conversions

The conversion of privately rented houses into flats for sale has been one of the major forms of new dwelling provision in Inner London in the 1980s and 1990s (Hamnett, 1989a; Barlow, 1989; London Research Centre, 1992). It began in the 1970s as rising prices pushed single-family home ownership out of reach in the more attractive areas of Inner London and developers saw the potential profits to be made from converting three- or four-storey terraced houses into flats, typically one per floor, but sometimes more in the larger terrace houses. The process was first encouraged by availability of government housing improvement grants which were given on a per-unit-created basis and which were open to landlords and developers as well as owners (Hamnett, 1973). Consequently, many developers quickly saw the potential to undertake government-aided conversions of privately rented houses into flats for sale. Demographic changes, particularly the rapid growth of one- and two-person households in London, increased the potential market for conversions.

During the 1980s and 1990s the phenomenon spread out widely across much of Inner London to anywhere where houses offer the potential for

Figure 6.5 (opposite) Percentage of private housing stock, 1981: (a) single-family dwellings, (b) purpose-built flats, and (c) conversions.
Source: Census of Population, 1981.

profitable conversion into flats. Conversions now form the first step on the owner-occupied housing ladder for many first-time buyers in Inner London, and luxury conversions of houses in prestige areas such as Hampstead are highly sought after. The LRC report estimated that the number of conversions in Greater London from 1980 to 1989 totalled 120,700 units, made up of 40,900 in central London, 43,100 in the rest of Inner London and 36,700 in Outer London. This represented a net gain of some 80,000 units. In the same period, around 135,000 new-build dwellings were provided in all tenures. Not surprisingly, the report concluded that 'Conversions are therefore a major source of additional accommodation in London.' In many areas of Inner London flat conversions now represent the dominant form of owner-occupied accommodation and the number is likely to have grown substantially during the 1990s.

Conversion of warehouses, offices and industrial buildings

The conversion of former industrial or warehouse property into flats began in London in the 1980s with the conversion of some of the old derelict riverside warehouses near St Katherine's Dock and Wapping. Warehouse conversions and associated new-build apartments spread rapidly along the riverside east of Tower Bridge before the market spectacularly collapsed in 1989 with the housing market slump. The process got under way again very rapidly from the mid-1990s onwards when the home ownership market picked up, and almost every old riverside warehouse capable of conversion has been converted. They offer the advantage of proximity to the river and to the City of London, as well as often spectacular internal spaces.

The 1990s saw the expansion of conversions away from the river to former industrial and office properties in the City fringe areas such as Clerkenwell and Shoreditch and more generally across central London. This took place for two main reasons. First, and most important, the sharp downturn in commercial office rents and property values, particularly in many marginal areas, during the recession of the early 1990s led to a wave of bankruptcies and distress sales which prompted both owners and developers to look for ways of increasing the return on their properties. The successful import of the loft-living idea from Soho in New York by one or two pioneering developers such as Harry Handelsmann's Manhattan Loft Corporation convinced others that the market was feasible (Zukin, 1998). Second, the loft-living concept became very fashionable in avant-garde circles in the late 1990s, prompted by very clever marketing, offering the possibility of large open-plan living spaces in stylish old buildings in the heart of London (Hamnett and Whitelegg, 2001). In addition, the conversion of warehouses in Docklands had shown that there was a large market for such properties, particularly in close proximity to the City which allowed financial and other workers to get to work easily without a long commute. Many of the converted buildings have been featured in design magazines, and some districts such as Clerkenwell and Shoreditch have been promoted as the hot, new, desirable residential areas of London, with designer bars, clubs and restaurants all contributing to the changes in social composition.

The scale and extent of these conversions has been dramatic. ESRC-funded research by Hamnett and Whitelegg (2001) has shown that in Clerkenwell alone almost three hundred planning applications were made for change of use to residential between 1991 and 1998, involving 2,700 units. The number of applications increased dramatically in the late 1990s as the popularity of loft-living became clear and developers vied to convert buildings. While an increasing proportion of these units were small apartments rather than lofts *per se*, loft-living is clearly here to stay in Inner London. Marsh (1999) has shown that around 10,000 units were added to the housing stock in central and Inner London in the 1990s as a result of office, warehouse and factory conversions.

The sale of local authority homes for ownership

The sale of local authority or council property for owner-occupation began in 1980 with the introduction of the Conservatives' 'right-to-buy' legislation which gave council tenants a right to purchase their homes at a discount if they had lived there for a minimum period of years. The impact of RTB has varied, however, depending both on the nature of the housing stock and the attitude of local authorities. In areas where the majority of council property is made up of high-rise and high-density blocks RTB sales have generally been limited, whereas in areas of more attractive terraced or semi-detached houses, sales have been much faster (Dunn *et al.*, 1987; Forrest and Murie, 1988; Kleinman and Whitehead, 1987). As a consequence, RTB sales have generally been more marked in Outer London boroughs than in Inner London. In addition, sales in authorities such as Camden, which opposed RTB legislation, were relatively limited, at least initially, whereas in some Conservative authorities, notably Westminster and Wandsworth, RTB sales were pushed with an enthusiasm bordering on the evangelical. In these boroughs sales were part of a broader strategy to reshape the resident composition socially and politically into a safe Conservative electorate by fostering home ownership under the guise of 'building stable communities'.

In Westminster, led by Dame Shirley Porter, the Conservative council's large-scale selective housing sale policy was investigated by the district auditor on the basis that it may have been politically designed to reduce the number of Labour-voting council tenants in some of the more electorally marginal wards and to replace them with homeowners who were thought to be more likely to vote for the Conservatives. Homeless families were also to be rehoused outside the borough.The results of the investigation were damning. In 1994 the district auditor, an accountant with the firm of Touche Ross, produced a report which stated that:

> no local authority acting reasonably would adopt a policy, the anticip-
> ated consequences of which were that the numbers of homeless house-
> holds placed in temporary accommodation would increase by 172 each
> year and that the net revenue cost to the Council would increase, as a
> result, by over £1.5 million in each succeeding year on the previous
> year's costs. In my provisional view, it was a decision no local housing
> authority acting reasonably would have taken given the unanticipated
> financial consequences, having regard to its fiduciary duty to local
> taxpayers.
>
> (Touche Ross, 1994, 22.d.ii)

The report further found that the council had been guilty of gerrymandering; that is to say, the manipulation of electoral boundaries and/or the
composition of the electorate in order to win or retain political power. It
stated that:

> both the decision to increase the number of designated sales and the
> selection of the properties designated for sale were influenced by an
> irrelevant consideration, namely the electoral advantage of the majority
> Party. I am minded to find as a fact that the electoral advantage of the
> majority Party was the driving force behind the policy of increased
> designated sales and that that consideration was the predominant con-
> sideration which influenced both the decision to increase designated
> sales by 500 per annum and the selection of properties designated for
> sale. My provisional view is that the Council was engaged in gerrymander-
> ing, which I am minded to find is a disgraceful and improper purpose,
> and not a purpose for which a local authority may act.
>
> (Touche Ross, 1994, 22.e)

The district auditor's conclusion was that the council had sold homes improperly, and that six councillors and officers were collectively liable for a
sum of £31.7 million which he believed had been lost as a result of the
policy. The press coverage of the report was predictably critical:

> 'Housing policy was disgraceful gerrymandering' (*The Times*, 1994)
> 'Westminster housing rhetoric echoed Mao' (*Independent*, 16 October
> 1994)
> 'Westminster hid lost millions' (*Independent*, 1 February 1995)
> 'Westminster scandal a nation's shame' (*Observer*, 26 March 1995)
> 'How fear of defeat led to Tory housing scandal' (*The Times*, 10 May
> 1996)

Dame Shirley Porter and her colleagues unsuccessfully appealed to the High Court against the district auditor's judgement, and in 1997 the High Court found that she was guilty of 'wilful misconduct' (*Independent*, 1997). She successfully appealed to the Court of Appeal in 1997 against the surcharge, but in December 2001 the House of Lords unanimously reinstated the original decision which found that Lady Porter was guilty of political corruption and of a 'deliberate, blatant and dishonest misuse of public power' (Gibb, 2001a; Waugh, 2001).

Westminster was not alone. Wandsworth, the other Tory flagship council in Inner London, also initiated a strong policy of council house sales and direct disposals. Usher (1987) gives the example of the St John's estate in Battersea, which was sold by Wandsworth to Regalion Properties. The estate (now renamed Battersea Village) consisted of 284 flatted dwellings built in the 1930s and it was bought by Regalion for £3.7 million or £13,640 per flat. The flats were subsequently refurbished at a cost of £20,000 per unit, and then sold for a total profit of £14 million. Wandsworth was also found guilty by its district auditor of acting improperly in selling council houses without taking into account its statutory duties to the homeless. He found 'a relatively high correlation' between the level of housing expenditure and the five most politically marginal wards in Wandsworth between 1987–8 and 1990–1, the period when the Conservatives turned a single-seat majority into a majority of 35 in 1990. But in Wandsworth, the Conservatives were not accused of gerrymandering by seeking to sell properties on estates in marginal wards. Instead, Tory policy seems to have been aimed at changing the political makeup of the whole borough through a policy of housing-market-led, large-scale social engineering which was admitted by the council leader. Forrest and Murie, who term the process 'municipally managed gentrification', comment that:

> New forms of privatisation are . . . creating new pressures for inner London council tenants. The high valuations and prices which act against individual purchases create the ideal conditions for the transformation of former council estates into exclusive, executive housing for London's expanding, high earning professional classes.
>
> (Forrest and Murie, 1988, p. 146)

Socio-tenurial polarisation

The decline of private renting and the emergence of council renting and ownership in the last forty years is associated with a growing social division between the two main tenures. Private renting was, partly by virtue of its very size, a heterogeneous tenure which embraced the whole range of social groups from wealthy upper middle classes to unskilled workers, though the different groups lived in different areas and in very different housing conditions. Initially, council housing catered to the better-off skilled manual and junior white-collar workers. It was a desirable and scarce commodity and families competed to gain access. This is clearly shown by the social composition of former London County Council Housing which in 1938 was overwhelmingly (80 per cent) occupied by skilled, semi-skilled workers and junior manual workers. The unskilled accounted for only 20 per cent of the residents. This continued into the 1960s and it was only in the 1970s that the social composition of council housing began to change as the skilled manual workers began to move into home ownership and council housing opened up to the less skilled, the unemployed, ethnic minorities and single parents who would hitherto have been largely excluded on various grounds. The social gap between the two tenures then began to open up in a dramatic fashion, as Forrest and Murie (1988), Bentham (1986), Hamnett (1984) and others have shown. Council renting has ceased to be a desirable tenure for many skilled and junior white-collar workers who have either moved into ownership or moved out of Inner London, and it has become home to many of those who, thirty years ago, were effectively excluded from council housing. In the process it has become increasingly socially residualised. Those who can now buy, and those who cannot must rent. As this process of polarisation has proceeded it has fed upon itself, making social renting progressively less attractive as an option to those with any degree of effective choice. Those who are now left in the council sector in much of Inner London generally lack effective choice and include the unskilled, the unemployed, the economically inactive, single parents, refugees and some ethnic minority groups (Watt, 2001).

Contemporary trends in tenure polarisation can be examined using indicators such as income, class, unemployment or the percentage of households receiving supplementary benefits in each of the main tenures (Forrest and Murie, 1990; Hamnett and Randolph, 1987). The analysis below uses the data on social class and tenure in the decennial censuses. Although the definitions of social class vary somewhat from one census to another, the 1971, 1981 and 1991 censuses provide a clear overall picture. It is possible to compare the distribution of socio-economic groups by tenure in 1971 and 1991. Not surprisingly, the professional and managerial groups are strongly represented in home ownership and under-represented in the council sector, whereas the less-skilled groups are under-represented in home ownership and strongly over-represented in the council and housing association sectors. It is also clear that the magnitude of this division has increased over time.

Table 6.1 shows that in 1971 white-collar workers made up 41 per cent of the home ownership sector in Inner London and skilled manual workers accounted for 22 per cent. Just over 22 per cent of owners were economically inactive. In contrast, white-collar workers accounted for 17 per cent of the local authority sector, where the largest groups were skilled manual workers (27 per cent) and the economically inactive (29 per cent). By 1991

Table 6.1 Tenure by socio-economic group, Inner London, 1971, 1981, 1991

	Percentage owner-occupation			Percentage council housing		
	1971	1981	1991	1971	1981	1991
Prof. and managerial	22.1	27.3	29.2	4.2	4.9	4
Other non-manual	18.7	25.7	25	13.3	14.7	12
Skilled manual and own account workers	22	18.4	12.8	26.9	22	10.3
Semi-skilled	9.3	9.6	6.1	14.4	18.4	10.4
Unskilled	4.2	2.9	1.9	10.7	9.5	5
Armed Forces/OID*	1.2	2.2	1.3	1.7	3.8	1.1
Total econ. active	77.5	86.1	76.3	71.2	73.3	42.8
Economically inactive	22.5	13.9	23.7	28.8	26.7	57.2
Total	100	100	100	100	100	100

Sources: Censuses of Population, 1971, 1981 and 1991.

Note: * OID = Occupation inadequately described.

the proportion of white-collar workers in ownership rose slightly to 54 per cent, the economically inactive remained stable at 24 per cent, and all the groups of manual workers declined from 35 per cent to 21 per cent of the total. But in the council sector, the economically active groups fell dramatically from 73 per cent to 43 per cent, while the proportion of the economically inactive rose from 27 per cent to 57 per cent. This is a remarkable statistic which overshadows almost all others. The council sector is now dominated by those who are not in the labour force, by virtue of either long-term unemployment, retirement, ill-health, disability or single parenthood. This is clear evidence of residualisation of the council sector. The housing association sector runs it a close second with 49 per cent of households listed as economically inactive. The contrast with home ownership is very sharp.

The development of socio-tenurial polarisation can also be shown in a different way by looking at the changing tenure composition of different socio-economic groups. Table 6.2 shows the proportion of different socio-economic groups in different tenures in Inner London in 1981 and 1991. The proportion of households in owner-occupation increased from 27.4 per cent to 40.7 per cent, while the proportion in council housing remained stable around 43 per cent and the proportion in private renting halved from 30 per cent to 16 per cent. What is important, however, are the tenure distinctions by socio-economic groups. In 1981 62 per cent of professionals owned in Inner London, while 31 per cent rented privately and just 6 per cent rented from a local authority. The proportion owning falls steadily by socio-economic group, from 50 per cent of managers to just 12.6 per cent of unskilled. Conversely, the proportion renting from a local authority rises steadily from 6 per cent of professionals to almost two-thirds (65 per cent)

Table 6.2 Socio-tenurial polarisation, Inner London, 1981–91

SEGS	1981 (%)			1991 (%)			Change, 81–91 (%)		
	Own	PR	LA	Own	PR	LA	Own	PR	LA
3, 4	62.4	31.2	6.4	72.7	20.2	7.1	10.3	−11.0	0.7
1, 2, 13	50.3	31.7	17.9	67.9	18.6	13.5	17.5	−13.1	−4.4
5, 6	32.8	37.9	29.3	50.4	20.9	28.7	17.6	−17.0	−0.5
8, 9, 12, 14	26.7	23.6	49.7	47.0	12.6	40.4	20.3	−11.0	−9.3
7, 10, 15	17.9	28.6	53.5	30.6	14.7	54.7	12.8	−13.9	1.2
11	12.6	22.8	64.6	23.9	12.4	63.7	11.3	−10.4	−0.9
Total econ. active	30.1	30.1	39.8	50.8	17.6	31.6	20.8	−12.5	−8.3
Econ. inactive	17.8	28.9	53.3	24.9	13.2	62.0	7.1	−15.8	8.7
Total	27.4	29.9	42.7	40.7	15.9	43.4	13.3	−14.0	0.7

of the unskilled. Looking at the economically inactive, just 18 per cent owned (predominantly retired owners) while 53 per cent rented from a local authority. Comparing these figures to those for 1991, the proportion owning rose in every socio-economic group, to 73 per cent of professionals and 24 per cent of unskilled workers. In percentage point terms, the greatest increases were among managers, junior white-collar workers and skilled manual workers. The key point is that whereas 53 per cent of the economically inactive in Inner London were in council housing in 1981 this had risen to 62 per cent in 1991. The council sector is now increasingly home to those outside the labour market. This is shown visually in Figure 6.6.

House prices and affordability

One of the most dramatic effects of the change in London's occupational and incomes structure and the decline of private renting has been the rise in demand for ownership and the associated rise in prices. This has been strongly driven by the growth in the number of high-income professional and managerial households working in the City and the creative industries. With salaries and bonuses running into millions of pounds at the top, there has been very strong demand for housing at the top end of the market. In addition there has been a continuing demand from the international rich, including film stars and musicians such as Madonna who bought a £7.5 million house in Notting Hill (*Observer*, 2000) and Robert de Niro (*The Times*, 2001). Given the limited supply of top properties, this demand has percolated downwards. A graduate recruit at one of the top City law firms or merchant banks on a salary of £50,000 can get a mortgage of over £150,000 (Thornton, 2000a).

These changes are shown by the Land Registry figures on house prices by borough. In June 1995, when London was slowly emerging from the

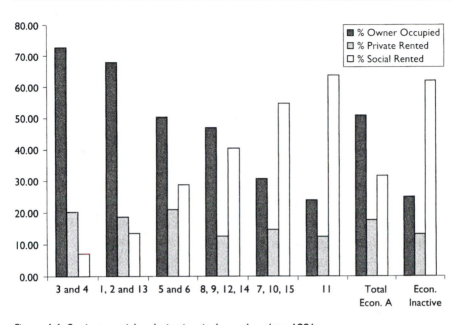

Figure 6.6 Socio-tenurial polarisation in Inner London, 1991.
Source: Census of Population, 1991.

housing market slump of the early 1990s, the average house price stood at
£98,300. In June 2000, the average price was £193,000: an increase of 96 per
cent in just five years. What is particularly significant, however, is that
percentage increases were generally much higher in the central and Inner
London boroughs than they were in Outer London. At the top end both
Tower Hamlets (Docklands) and Kensington recorded increases of over
150 per cent. All the other boroughs recording increases of over 100 per cent
were in Inner London, with the exception of Kingston. Lambeth recorded
an increase of 130 per cent and Hackney of 116 per cent (Table 6.3 and Figure 6.7). At the other end of the scale, several peripheral London boroughs,
such as Bromley, Bexley and Barking, recorded increases of under 70 per
cent, with Barking, the most easterly working-class borough, recording an
increase of just 55 per cent: only one-third of that in Docklands. This points
both to the importance of centrality for housing demand and to the extent
of gentrification in Inner London. The correlation between the distribution of house price changes by borough in the second half of the 1990s and
the geography of professionalisation in the 1980s is remarkably strong – a
point which is developed in more detail in Chapter 7 on gentrification.
  Paralleling the growth of house prices in London, there has also been a
sharp rise in the number of properties at the top end of the market changing
hands for prices of over £1 million. Land Registry data for April–June 2000

Table 6.3 Percentage mean house price change in London, by borough, 1995–2000

	April–June 1995	April–June 2000	Absolute 1995–2000	% Change 1995–2000
Tower Hamlets	76,168	198,851	122,683	161.1
Kensington	218,995	553,782	334,787	152.9
Lambeth	76,847	176,494	99,647	129.7
Westminster	181,375	404,086	222,711	122.8
Hackney	71,060	153,871	82,811	116.5
Hammersmith	133,464	278,036	144,572	108.3
Kingston	94,295	195,043	100,748	106.8
Camden	148,412	305,631	157,219	105.9
Wandsworth	110,583	227,104	116,521	105.4
Islington	118,761	241,078	122,317	103.0
Haringey	85,409	172,158	86,749	101.6
Merton	93,995	187,709	93,714	99.7
Brent	77,645	153,461	75,816	97.6
Richmond	143,128	282,554	139,426	97.4
Ealing	93,554	183,642	90,088	96.3
Southwark	84,154	164,970	80,816	96.0
Hounslow	88,673	172,237	83,564	94.2
Lewisham	60,012	114,082	54,070	90.1
Barnet	106,571	201,278	94,707	88.9
Newham	48,305	91,188	42,883	88.8
Harrow	88,596	166,717	78,121	88.2
Waltham Forest	56,463	103,796	47,333	83.8
Greenwich	70,664	126,784	56,120	79.4
Croydon	72,090	127,891	55,801	77.4
Redbridge	73,872	130,234	56,362	76.3
Sutton	78,708	136,771	58,063	73.8
Enfield	76,613	131,654	55,041	71.8
Bromley	96,214	160,741	64,527	67.1
Hillingdon	83,008	137,681	54,673	65.9
Havering	74,233	119,822	45,589	61.4
Bexley	66,178	106,533	40,355	61.0
Barking	50,378	78,157	27,779	55.1
Greater London	98,300	193,004	94,704	96.3

Source: Land Registry Statistics.

show that there were a total of 107 properties in England and Wales which sold for over £2 million. Of these, no less than 90 (84 per cent) were in Greater London. A total of 394 properties sold for over £1 million in England and Wales, of which 293 or 74 per cent were in Greater London. Of the properties of over £1 million, 87 (30 per cent) were in Westminster, and 112 (38 per cent) were in Kensington and Chelsea, with the rest scattered across the remaining high-status boroughs (Table 6.4). This concentration of high-value sales is indicative of the impact of the growing number of high

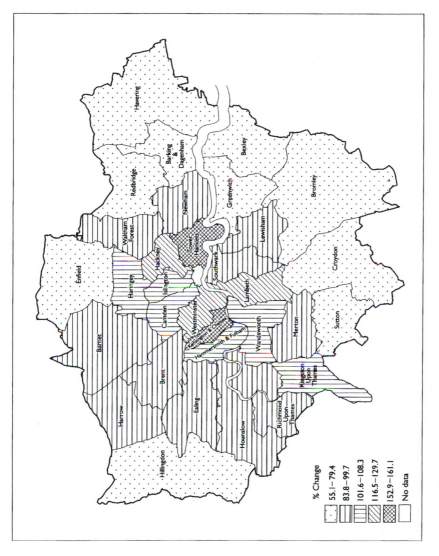

% Change
55.1–79.4
83.8–99.7
101.6–108.3
116.5–129.7
152.9–161.1
No data

Figure 6.7 The distribution of average property price changes, by borough, 1995–2000.
Source: Derived from calculations from Land Registry statistics.

Table 6.4 Property sales of £1 million and over, London boroughs, July–September 2000

	£1m–1.5m	£1.5m–2m	>£2m	Total	% of total
Kensington	43	29	33	105	31.8
Westminster	39	12	17	68	20.6
Wandsworth	33	8	3	44	13.3
Camden	14	3	4	21	6.4
Richmond	12	4	5	21	6.4
Barnet	6	5	2	13	3.9
Merton	4	4	3	11	3.3
Hammersmith	6	1	1	8	2.4
Hounslow	5	3	0	8	2.4
Islington	6	1	0	7	2.1
Bromley	3	2	0	5	1.5
Haringey	1	1	2	4	1.2
City of London	3	0	0	3	0.9
Greenwich	1	1	1	3	0.9
Tower Hamlets	1	0	2	3	0.9
Ealing	2	0	0	2	0.6
Enfield	1	1	0	2	0.6
Harrow	1	0	0	1	0.3
Sutton	1	0	0	1	0.3
Greater London	182	75	73	330	100.0

Source: Land Registry Statistics, July–September 2000.

earners and wealthy individuals, many from overseas, in the London housing market (Thornton, 2000a). City financial workers are particularly important in this market and it has been estimated that perhaps 50 per cent of annual city bonuses are channelled into the housing market in London and the South East. The sharp cut-backs in City bonuses noted in Chapter 4 are therefore likely to have a marked impact on the top end of the London housing market.

It is also illuminating to look at the average prices by borough. The average price in London as a whole in June 2000 was £193,000, but in Kensington and Chelsea, the most expensive borough, the average price was £553,000, followed by Westminster at £404,000. The average price for detached houses in Kensington and Chelsea was £2.89 million, with semi-detached houses fetching £1.35 million and terraced houses £1.25 million. Even flats, the most common property type, sold for £380,000 on average.

At the other end of the scale, there were only two boroughs in June 2000 with average prices of under £100,000, both of which are in East London: Barking and Dagenham with an average price of £78,000, and Newham with an average of £92,000 (Table 6.3). From June 2000 to 2001 the average price in Barking and Newham rose by 16 per cent and 21 per cent respectively, and similar increases were also seen in the other low-price boroughs. The most recent Land Registry figures to March 2002 once again show the lowest price boroughs recording large increases.

Housing affordability in London

High prices in London have led to a major housing affordability problem. Given that the average gross annual household income in London from 1996 to 1999 was £27,000 (ONS, 2001) this indicates that households on average income could only afford to buy the average property in the lowest-price boroughs and only with a mortgage approaching four times household income. Calculations based on averages are of course very crude, and the Land Registry data show that the average price of flats and maisonettes in the lower-priced suburban boroughs in June 2001 ranged from £65,000 in Barking and Dagenham to £74,000 in Bexley and £78,000 in Havering. Given that these prices are averages, the price of a one-bedroom flat would be lower, and would be affordable to a household on average income. But, as no less than 56 per cent of all households had average incomes of less than £23,400 p.a., many of them will be effectively excluded from home ownership even in the cheapest borough. Generally speaking, first-time buyers have lower incomes than older and more established households, which will also intensify the difficulties. The problem of housing affordability for lower-income workers was recognised in a Greater London Authority report (2000) which recommends, amongst other things, a rise in the proportion of affordable housing constructed under section 106 agreements which compel developers of schemes of fifteen houses or more to set aside a specific proportion of dwellings or to give the local authority a cash equivalent for low-cost housing.

The housing affordability problem in London is not new. It has been a constant since the beginning of the 1970s when house prices first began to rise steeply in London. In the first house-price boom in the period 1971–3 house prices doubled from £5,000 to £10,000, and they doubled again in the second boom in the late 1970s (Hamnett, 1999). They more than doubled during the boom of the mid–late 1980s, and, although they fell back sharply in the slump of the first half of the 1990s, they have subsequently more than doubled since 1995, from £98,000 in June 1995 to £206,000 in June 2001. Although prices have also risen sharply in England and Wales as a whole, average prices in London are now some 76 per cent higher than in England and Wales as a whole. The gap between average house prices in London and the South East and the rest of Britain has widened substantially in recent years.

The argument made here is that this is a direct reflection of the growing pressure of high-income demand, particularly from affluent professional and managerial dual-earner households, combined with a relatively fixed supply of property. As a result, pressure on prices has pushed its way down from the top end of the market where the demand for desirable properties is perhaps most intense, in a spillover process which progressively increases prices in many of the less attractive areas as potential buyers are forced further downmarket in the search for an attractive yet affordable home. At the same time, the growing number of young professionals seeking to buy is putting pressure on the lower end of the market. Needless to say, the groups squeezed out are the lower- and middle-income households who find it almost impossible to purchase. It is important to bear in mind here that earnings and incomes in London have risen far more at the top end, which increases demand pressure.

The 1991 census evidence shows that there are a substantial number of less skilled homeowners, but it can be argued that many, if not most, of these are those who were fortunate enough to buy thirty or more years ago when prices were substantially lower. As a consequence, the social mix found in many Inner London areas can be seen to be a result of the historical legacy of successive waves of buyers, with older households now living in areas they could no longer afford if they were buying today. As they die or move they will be gradually replaced with households able to pay current prices. It is only in some of the cheaper, more peripheral, boroughs, that middle-income groups can gain a foothold in the market. Consequently, teachers, social workers, policemen and women, nurses and other groups of key workers are increasingly unable to afford to live in Inner London (Greater London Authority, 2001). This problem has been compounded by the sale of police houses and nursing homes in Inner London by the Metropolitan Police and by National Health Trusts. In addition, there is a large problem of both homelessness and of temporary housing in bed and breakfast accommodation (Bramley *et al.*, 1988).

Demographic change and housing supply

The problem of housing costs and affordability is likely to be intensified by the fact that London's population, which fell steadily from its peak in 1939 to a low of 6.8 million in 1983, has now begun to increase again as a result of natural increase and net in-migration and is currently estimated to be 7.2 million. Net international migration is estimated to have added 120,000 in 1999/2000 against a loss of 66,000 to the East and South East region. In addition, household numbers are continuing to rise. There are currently some 3.1 million households in London (up from 2.84 million in 1991), but it is estimated that this could rise to 3.6 million by 2016: an increase of 31,300 households a year. This increase in household numbers is taking place against a long-term fall in the volume of new housing completions in London. In the late 1960s and mid-1970s, 30,000 new dwellings were built each year, of which 25,000 were social housing. This has subsequently fallen

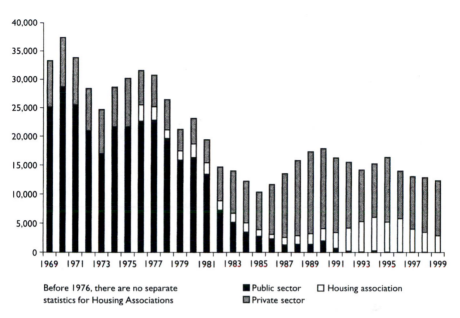

Figure 6.8 Completions of new housing in London, 1969–99.
Source: Greater London Authority: data taken from *Housing and Construction Statistics, Great Britain*, various years.

sharply and in recent years only about 12,000 units were built each year of which 3,000 were social housing (Greater London Authority, 2000, 2002). When conversions are included there are about 19,000 net extra dwellings per year against an estimated housing demand of 43,000. The result will be a growing housing shortage, which will particularly affect low-income households unable to compete in the ownership market (Figure 6.8).

Conclusions

The transformation of London's occupational structure and the associated changes in the structure of earnings and incomes have generated a large and growing number of high-income professional and managerial workers. Given the sharp decline of private renting until the 1990s and the financial attractions of home ownership, the demand for ownership has been intensified. The supply of housing has grown only slowly, however, and the result has been a sharp rise in prices – much higher than in Britain as a whole. There has also been a displacement of demand from the traditional middle-class residential areas as prices have risen to stratospheric levels. In the process what remains of the working class is being gradually replaced in traditional working-class areas as middle-class owners move in. Home ownership in

Inner London is increasingly becoming the preserve of the well off, as Ruth Glass perceptively predicted over thirty years ago.

The decline of the private rented sector has been accompanied by the growth of home ownership and council renting. The transformation from renting to owning has taken a wide variety of forms, including conversion of rented houses into flats and conversion of industrial and commercial property to residential. The council sector peaked in 1981 and since then it has been in long-term decline owing to an almost complete fall off in new building, right-to-buy sales and stock transfers to housing associations. Although some of the worst high-rise blocks have now been demolished, much of the remaining stock consists of high-rise or high-density flats. This housing now provides the major source of accommodation for many of the less skilled, the unemployed, lone-parent families and the economically inactive. Thus, as Gillian Tindall foresaw thirty years ago, a growing divide has opened up in London between home ownership and the social rented sector which has become increasingly residualised. Although this trend is not unique to London, its effects are intensified as a result of the very high price of owner-occupation.

Chapter 7

Gentrification and the middle-class remaking of Inner London

London is now being 'renewed' at a rapid pace – but not on the model about which we are so often warned. Inner London is not being 'Americanised': it is not on the way to becoming mainly a working class city, a 'polarised' city, or a vast ghetto for a black proletariat. The real risk for Inner London is that it might well be gentrified with a vengeance, and be almost exclusively reserved for selected higher class strata.

(Glass, 1973)

London is being 'made over' by an urban centred middle class. In the post war era, upwardly mobile social classes tended to leave the city. Now, led by a new middle class they are reconstructing much of inner London as a place both in which to work and live.

(Butler, 1999, p. 84)

Introduction

The transformation of London's industrial, occupational and earnings structure and the continuing growth of the professional, managerial and technical middle class has placed enormous pressure on London's housing market over the last thirty years. Not only has home ownership grown considerably, but the middle classes have expanded and pushed into what were traditionally working-class areas of the city. This process of middle-class residential expansion into hitherto working-class areas is commonly termed 'gentrification', and it has dramatically reshaped the housing market and the social geography of Inner London as well as a number of other major western cities. The term was coined by Ruth Glass, an émigré sociologist who came to Britain from Germany in the 1930s. She was intrigued by the social and housing changes which she saw taking place in Inner London in the 1950s, and she perceptively commented in the introduction to her book *London: Aspects of Change*, published in 1963, that:

One by one, many of the working class quarters of London have been invaded by the middle classes – upper and lower – shabby modest mews and cottages – two rooms up and two down – have been taken over when their leases expired, and have become elegant, expensive residences. Larger Victorian houses, downgraded in an earlier or recent period – which were used as lodging houses or were otherwise in multiple occupation – have been upgraded once again . . . Once this process of 'gentrification' starts in a district it goes on rapidly until all or most of the original working class occupiers are displaced and the whole social character of the district is changed.

(Glass, 1963, p. xviii)

Her use of the term 'gentrification', which sometimes vexes American academics who would prefer the term neighbourhood 'revitalisation' or 'renovation' was deliberately ironic and rooted in the intricacies of traditional English rural class structures. It was designed to point to the emergence of a new 'urban gentry', paralleling the eighteenth- and nineteenth-century rural gentry familiar to readers of Jane Austen, who comprised the class strata below the landed aristocracy, but above yeoman farmers and landless labourers. She identified gentrification as a complex process involving physical improvement of the housing stock, housing tenure changes from renting to owning, price rises and the displacement or replacement of the working-class population by the new middle class. She was not alone in recognising the emergence of gentrification in London. Several key elements of the process were brilliantly captured by Michael Frayn (1967) in his novel, *Towards the End of the Morning*:

They decided to find a cheap Georgian or Regency house in some down at heel district near the core. However depressed the district, if it was Georgian or Regency, and reasonably central, it would soon be colonized by the middle classes. In this way they would secure an attractive and potentially fashionable house in the heart of London, at a price they could afford; be given credit by their friends for going to live among the working classes; acquire very shortly congenial middle class neighbours of a similarly adventurous and intellectual outlook to themselves; and see their investment undergo a satisfactory and reassuring rise in the process.

(Frayn, 1976)

The complex cultural and aesthetic dimensions of the gentrification process involving the physical renovation of older houses to fit the standards, tastes and lifestyles of the new owner-occupying middle-class residents have been discussed by a variety of academic commentators, notably Jager (1986), Ley (1996) and Butler (1997), but one of the first writers to capture some of the cultural manifestations of gentrification in Inner London in the early 1970s was Jonathan Raban (1974) in his book *Soft City*:

The first sign is a crisp white painted house front. Outside, one of those continental biscuit-tin cars, a Renault 4L or a Citroën 2CV is parked. Inside, through the window – it has blinds not curtains – one spots a Japanese paper lampshade, a smart little bookcase of the kind you get on mailorder through the *Observer*, stacked with glossy volumes of reproductions, a stripped pine table, a long sealed and sanded floor with dead sheep for carpets. The middle interior wall has gone, and one can see through into the back garden with its breakfast patio. The knockers-through are here.

(Raban, 1974)

As someone who frequently wandered the streets of Primrose Hill, Camden Town, Kentish Town, Islington and Notting Hill in north London in the early 1970s, looking at the builders' skips and renovation, I can testify to the accuracy of Raban's depiction. Growing numbers of houses in these areas showed the visual signs of gentrification as decaying, poorly maintained private rented shared housing gave way to middle-class, single-family owner-occupation and associated renovation (Figure 7.1).

Gentrification and social change in Inner London

The changes which Glass, Frayn and Raban identified in Inner London in the 1960s and early 1970s have subsequently grown considerably in importance. There has been an increase in the number and proportion of professional and managerial workers and a transformation in the tenure structure and quality of the housing stock in many parts of Inner London. Initially, gentrified areas were found in parts of Chelsea in the 1950s and then in Barnsbury and some of the Islington Squares in the 1960s and early 1970s, but they rapidly spread across much of inner northern and western London, and by the 1990s they had spread rapidly into some of the last bastions of working-class eastern and southern Inner London: areas such as Hackney, Stoke Newington and Brixton. In the process, the character and social mix of many of these areas has been dramatically transformed. They have gradually changed from down-at-heel areas of working-class, private rented housing to smart areas of middle-class owner-occupation complete with accompanying bars, restaurants, upmarket food and designer shops and the all-pervasive estate agents. The plethora of estate agent offices are an important sign that the areas have been incorporated into London's burgeoning owner-occupied property market.

The importance of gentrification in dramatically reshaping the social geography of Inner London over the last 30–40 years cannot be overstated. During the nineteenth century, the population of London grew dramatically as a result of in-migration – from 1 million in 1801 to 6.5 million in 1901. Almost all of this growth was concentrated in what is now known as

Figure 7.1 The location of gentrified and gentrifying areas of Inner London.
Source: Chris Hamnett.

Inner London, the old London County Council area, and it was associated
with a rapid expansion of the working classes, and an increase in poverty,
deprivation, overcrowding and poor housing conditions. As Stedman
Jones (1971), Green (1986), Dyos (1967), Wohl (1971) and others have
shown, extensive areas of central and Inner London were transformed into
slum housing during the nineteenth century. As a consequence of these
pressures, the middle classes also abandoned early nineteenth-century suburbs such as Barnsbury in Islington, Primrose Hill, Camden Town, Bayswater
and Notting Hill in the face of growing working-class population pushing
out from the centre.

By the end of the 1950s, as a result of continued suburbanisation, many
areas of Inner London built in the nineteenth century had become generally
run-down and occupied by new in-migrants such as the Irish in Camden

Town and Kilburn, and the West Indians in Brixton and Notting Hill. As Raban graphically put it:

> A combination of class fear and (mid nineteenth century) railway engineering turned a vast stretch of residential London into a no-man's land . . . Camden Town, Holloway, Islington were abandoned to a hopelessly entrenched working class. It's only in the last decade or so that a new middle class, trendy and pioneer, have replaced these buffer areas, between the nobs and the mob of N1 and NW1.
>
> (Raban, 1974)

Raban nicely catches the fact that gentrification dramatically changed the geography of class in London. Charles Booth, the social reformer, writing of London south of the river, stated in 1901 that: 'the population is found to be poorer, ring by ring as the centre is approached, while at its centre there exists impenetrable mass of poverty' ([1901] 1982). Unlike the pre-industrial city, where the middle and upper classes valued central city locations, the industrial city in the nineteenth century and the early twentieth was strongly characterised by decentralisation of the middle classes to new suburbs and the dominance of working classes in the inner city. This pattern was noted by Engels (1969) in Manchester in the mid-1800s, by Burgess in Chicago in the 1920s, and by other observers (Ward, 1989).

London saw a marked continuation of this process in the first decade of the twentieth century and in the inter-war years as the owner-occupied suburbs rapidly expanded (Jackson, 1974). In the inter-war period the geographical area of London doubled, largely through construction of semi-detached housing. As a result, although there were extensive middle-class areas in the central areas of Westminster, Kensington and St Johns Wood, with outliers in Hampstead, Highgate and Richmond, much of Inner London in the 1950s and 1960s was dominated by a relatively large working-class population running from Hammersmith and Fulham and north Kensington in the west, Camden, Islington in the north, Hackney, Tower Hamlets and Newham to the east, and south of the river in Bermondsey, Lambeth and Southwark. To an extent, the middle classes in London had either voluntarily suburbanised or fled the inner city.

The social geography of London in 1966 was clearly shown by a cluster analysis of 1966 census data undertaken by the Greater London Council (Daly, 1971). The GLC took eleven selected census variables on occupation, tenure, household structure and mobility for each of Greater London's 7,000 wards and examined the extent to which they were statistically interrelated. They identified twelve different clusters of wards, each of which was characterised by a different set of economic and social attributes. They then mapped the distribution of wards according to clusters, giving each cluster a different colour code (Figure 7.2).

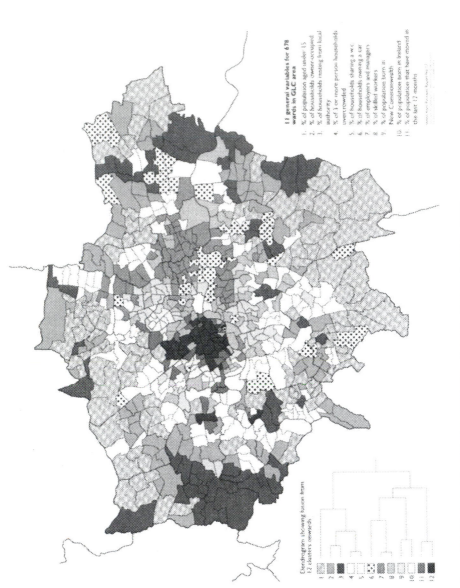

Figure 7.2 Clusters of social characteristics of wards, Greater London, 1966.
Source: Daly (1971).

Figure 7.2 shows a sharp contrast between the East End and the West End of London, with the former characterised by high levels of council housing and working-class residents, and the latter characterised by high levels of young, mobile, middle-class private renters. Surrounding central London, much of Inner London was characterised by working-class private rented housing. Then, further out, were the more middle-class, owner-occupied suburbs. What gentrification has done is to transform much of Inner London into an extension of the middle-class West End in social and tenure terms. In the process, what remains of the traditional working-class residents of Inner London has been squeezed into residual areas. Smith and Williams (1986) have termed this the victory of the 'space invaders'.

Who are the gentrifiers?

Gentrifiers conform to the characteristics first identified by Glass thirty years ago. They tend to be highly educated, highly skilled, and high-paid professional and managerial households, either childless or with young families. Hence the linkage of gentrification with yuppies (young urban professionals). These characteristics were also identified by Gale (1978) in his pioneering study of gentrification in the inner areas of American cities, and they have been confirmed in a number of studies of London (Munt, 1985; Bridge, 1994; Butler, 1997; Hall and Ogden, 1992). Butler found in Hackney:

> a group who (in comparison with the middle class as a whole) tend to come from middle class backgrounds, to have been educated in the private sector or at selective secondary schools, to have gone to Oxbridge or Redbrick universities and studied non-scientific or technical subjects and are now found working in professional as opposed to managerial or technical occupations. Where they are living in non-single person households, both partners almost always have full-time professional jobs and the household income is considerably higher than for a similar national group. Finally, and perhaps most tellingly, party political identification is biased towards the Labour party and support for the Conservative party is under-represented.
>
> (Butler, 1996, p. 93)

Not all gentrifiers conform precisely to this model. Rose (1984) identified what she termed 'marginal' gentrifiers, who may be in the lower ranks of the professional or technical occupations or work in low-paid creative jobs and have managed to pioneer gentrification in marginal areas. There also appears to be a sequence of gentrification, with the process being started by 'pioneers' who tend to be less well qualified, less established in their careers and lower paid, who need to find affordable housing in the city. By definition, those in this group seek out areas which have not yet been gentrified. As gentrification progresses, and prices rise, this group tends to be priced out of the market, and are replaced by a second wave who are more affluent and less risk-averse. Finally, when gentrification in an area becomes very firmly established, and prices have risen, a more highly skilled and highly paid group begin to move in. By this stage, the next generation of young pioneers have already moved on to another area. Most of Notting Hill, Clerkenwell and Islington would now fall into this category of well-established gentrification, Stoke Newington is in the second category, and areas like Hoxton are in the third. So well established is gentrification in Islington that there have allegedly been complaints made to the makers of Monopoly requesting that the Angel, Islington be taken out of its 'low rent' category and moved near the top of the board! (Mullins and Watson-Smyth, 2000).

The literature consistently shows that gentrification does not entail a 'return to the city' from the suburbs but rather involves young singles and couples, some of whom will have moved to London for higher education or on graduation and who often chose to remain in the city rather than moving out to the suburbs in their thirties. There is also a distinct gender dimension to the process as more women have gained degrees, entered professional, managerial or technical jobs and deferred childbearing. Warde (1991) suggests that 'An inner city location minimises journey to work costs for households containing more than one earner, enhances the efficiency of household production and facilitates substitution of marketed services for domestic ones' (p. 224). This is particularly important where dual-career professional households are involved who may work long or irregular hours and require relatively quick and easy access to their places of work.

One of the problems of this argument is that by no means all young professionals or managers choose to gentrify. The majority live in the suburbs or even further afield. There must therefore be some specific reasons why some households gentrify and the majority do not. The answer, according to Butler (1997), is that gentrifiers comprise a specific fraction of the professional and managerial middle classes, distinguished by high levels of education and a strong representation in the liberal arts. They tend to have a strong orientation towards the cultural and entertainment facilities offered by the central city and tend to be more to the left politically than the middle classes as a whole, who are generally more Conservative voting. This conforms to the argument put forward by Savage *et al.* (1988, 1992) regarding the spatial fragmentation of the 'service class' in contemporary Britain. The argument is that the managerial and technical fractions of the service class tend to have a different educational background, with a greater emphasis on scientific/technical subjects, a different set of social and cultural interests, stronger Conservative political affiliations and a different residential pattern.

Butler also suggests that different types of gentrified areas are also likely to contain different types of residents. He suggests that gentrifiers in Docklands or inner West London, which are characterised by a higher proportion of professionals working in finance and business services, are likely to be more conservative and materialistic in orientation than the 'post-1968' generation of political radicals he found in Hackney who were willing to live in socially and ethnically mixed inner-city areas (Crilley *et al.*, 1991; see Hall and Ogden, 1992). There is also likely to be a considerable difference in aesthetic orientations between gentrifiers who actively seek out eighteenth- and nineteenth-century houses or other buildings to renovate, and those who move into new purpose-built apartments in Docklands. The former group have an aesthetic commitment to period restoration and features, whereas the second group are perhaps more purely interested in the location. Interestingly, Butler and Robson state that in their research on gentrification in three areas of London, their hypothesis was that:

> different groups would be attracted to different areas and this would be determined by a range of factors, in addition to what they might be able to afford in particular housing markets. We believe that these divisions within the middle class are significant and reflect a broad range of 'cultural' and lifestyle factors and that they have a clear spatial manifestation.
>
> (Butler and Robson, 2001b, p. 2148)

The explanation of gentrification

The academic literature which tries to document and explain the rise of gentrification is very extensive (see Hamnett, 1984, 1991c; Smith, 1979, 1996; Butler, 1997 and Ley, 1996 for summaries), but it hinges around *three* key competing explanations. The first, put forward by David Ley (1981, 1996) and Hamnett (1984, 2000) and others, argues that the roots of gentrification lie in the changing industrial structure of major global cities with the switch from manufacturing industry to service-based industries and a concomitant change in the occupational class structure from one based around the dominance of a large manual working class to one increasingly dominated by white-collar professionals, managers and technical workers in the financial, cultural and service industries which are concentrated in major cities. Second, and related to this, it is also argued by Ley and Butler and others that as a result of these changes in class composition there have also been changes in cultural orientation and preferences and working patterns of a fraction of this new middle class which have predisposed them to living in the inner city, rather than commuting from suburbia. Bondi (1991), Warde (1991) and Butler and Hamnett (1994) have also pointed to the importance of changing gender relations, particularly the growing importance of dual-working professional households, which may necessitate closer proximity to work.

On the other hand, and strongly opposed to these interpretations, Smith (1979, 1987, 1996) has argued that the driving force behind gentrification is not the new middle class (whose existence he doubts), but the growing gap between property values and underlying land values in the inner city which have, he believes, opened up a growing 'rent gap' which has been exploited by the actions of property-based capital, estate agents, developers and the like, which has redeveloped or rehabilitated and gentrified undervalued inner-city housing for profit. In Smith's phrase, gentrification represents 'a back to the city movement by capital, not people'. In his view, the focus on cultural values and residential preferences of the new middle class is a diversion from the key issues which involve the supply of property and its financing, rather than demand and preference. As he put it:

> To explain gentrification according to the gentrifier's actions alone, while ignoring the role of builders, developers, landlords, mortgage lenders, government agencies, real estate agents, and tenants is excessively narrow. A broader theory of gentrification must take the role of producers as well as consumers into account, and when this is done, it appears that the needs of production – in particular the need to earn a profit – are a more decisive initiative behind gentrification than consumer preferences.
>
> (Smith, 1979, p. 540)

> The so-called urban renaissance has been stimulated more by economic than cultural forces. In the decision to rehabilitate an inner city structure, one consumer preference tends to stand out above the others – the preference for profit, or more accurately, a sound financial investment.
>
> (Smith, 1979, p. 540)

> A theory of gentrification must . . . explain why some neighbourhoods are profitable to redevelop while others are not. What are the conditions of profitability? Consumer sovereignty explanations took for granted availability of areas ripe for gentrification when this was precisely what has to be explained.
>
> (Smith, 1979, pp. 540–1)

Smith developed a sophisticated theoretical argument regarding the development of a 'rent gap' in the inner city which emerged as a consequence of major suburbanisation, and the devalorisation and abandonment of many American inner-city areas post-war. Smith argued that while the potential value of inner-city land remained high, the value of the buildings had fallen dramatically. Consequently, a growing 'rent gap' emerged between the potential value of the land and its existing use value. The size of the gap grew until it was possible for developers to move back to the inner city and profitably realise the underlying value of the land through renovation or redevelopment of the buildings. As Smith put it:

A theory of gentrification will need to explain the detailed historical mechanisms of capital depreciation in the inner city and the precise way in which this depreciation produces the possibility of profitable reinvestment. The crucial nexus here is the relationship between land value and property value.

(Smith, 1979, p. 542)

Smith is clearly right regarding the importance of capital depreciation and subsequent reinvestment in helping to explain gentrification, but his Achilles' heel is his inability or unwillingness to appreciate demand for inner-city locations from the middle class.

Smith (1987) accepts that 'western capitalist economies have experienced a decline in the relative importance of manufacturing employment and parallel increase in the importance of professional, administrative, service and managerial occupations in the producer services sector (finance, insurance, real estate and such), non-profit services (mainly health and education), and the government sector' (p. 153). He states that the 'transformation in occupational structures is undeniable' (p. 154) and he accepts that 'it is undeniable that professional, managerial and upper level administrative personnel in the expanding sectors noted above are heavily represented among gentrifiers' (p. 154).

But Smith takes as his indicator of class the Marxist position of relationship to the ownership and control of the means of production, and he argues that if there was a new middle class we would expect to see a redistribution of income towards the top of the income hierarchy. He analyses US income trends by quintile for the years 1947–78 and shows that there was very little change. He thus concludes that there is no evidence in income terms for the emergence of a new middle class. However, this analysis is flawed on two counts. First, the major redistribution of income in the USA to the higher-income groups has taken place since 1978, and second his analysis is at the level of the USA as a whole, rather than cities undergoing gentrification. But figures produced by Mollenkopf and Castells (1991) on income trends in New York, 1979–86, show that there has been a significant redistribution of incomes towards the top end, particularly the highest decile. These figures correspond to those for London, as Chapter 4 shows.

Smith's explanation is questionable for *five* main reasons. First, although it is true that much inner-city housing had become deteriorated and devalued in the post-war period and before, and this provided the necessary supply of devalued housing for the middle classes to occupy, Smith fails to explain why gentrification is concentrated in major cities with advanced service-based economies, and not in older manufacturing cities. Second, he fails to explain the basis of the demand for this renovated inner-city housing. Where have the growing army of middle-class buyers and renters come from, and why have they decided to buy in areas of the inner city they would not have even looked at a few decades before? Third, he fails to explain why gentrification occurs in some areas but not others, and fourth he fails to explain why gentrification began at a particular point in time but not before. Smith simply states that gentrification will take place when the rent gap is large enough. Finally, Smith is almost totally dismissive of the role of individual agency in the gentrification process. For him, it is a result of the decisions of capital to disinvest and reinvest, and demand is a misleading sideshow.

None of these arguments seem convincing, and the case made here is that the basis of an effective explanation has to rest on the demand side rather than the supply side of the equation. Indeed, there is a strong argument that while the role of capital may have been crucial in the gentrification process in the USA this is not the case in much of Britain where the principal agent of change has been the individual household who is seeking an affordable and potentially attractive place to live in the inner city. The role of professional residential developers in London has generally been confined to large-scale developments in Docklands, to conversion of commercial property into flats and the conversion of houses to flats (in all three their role has been extensive). As Butler and Robson note in their study of gentrification in three areas of London:

> We made a clear distinction between gentrification by collective action as opposed to capital. We argue that almost exclusively in London, the latter had been confined to the Docklands area ... The history of gentrification in London over the past nearly 40 years has been largely one of upgrading of mainly 19th century property by individuals or small scale developers.
>
> (Butler and Robson, 2001b, p. 2148)

Smith's objection to demand-led explanations is that they are overly individualistic, place too much stress on shifts in consumer choice and preference, and fail to provide an adequate explanation of underlying changes in the land and property markets. He also argues that they are insufficiently materialist in their theoretical approach in that they fail to deal with underlying economic changes. It is argued here, on the contrary, that a demand-based explanation which locates the basis of gentrification demand in the shifts in industrial, occupational and earnings structures which have taken place in the move from industrial to post-industrial cities is based on changes in underlying economic and class structures. The addition of arguments regarding the centrality of the tastes, cultural and political orientations of a specific fraction of the middle class does not replace the economic basis of the argument with one based on preferences: it supplements it with a cultural analysis of middle-class formation and reproduction. For a Marxist committed to analysis of class, Smith's (1987) unwillingness or inability to recognise the existence of an expanded urban middle class is disappointing.

Although a demand-based explanation must also account for the reasons why housing in the inner city had become deteriorated in the first place, my argument would be that this was a consequence both of a lack of effective demand from low-income, working-class renters and of the consequences of the denial of mortgage finance, as Williams has indicated. The growth of the new middle classes in a number of cities with growing financial and business service sectors, and the willingness for institutions to lend in inner-city areas, permitted gentrification to take off. There is also another explanation which has been put forward by Redfern (1997) which argues that gentrification took place because of the availability and falling real cost of domestic technologies which permitted old, nineteenth-century houses to be modernised and upgraded to twentieth-century standards. He argues that you cannot have gentrification without being able to modernise houses. Redfern has identified an important point here, but it seems to be a necessary rather than a sufficient factor for gentrification to occur. If the supply of gentrifiable properties and the demand for them was not there, domestic technology alone would be unlikely to lead to gentrification.

The geographical expansion of the professional and managerial classes

As Chapter 3 showed, London has experienced a dramatic increase in both the size and proportion of its professional and managerial labour force since 1961 as a result of the transformation of its economic base from manufacturing industry to services and the associated changes in its occupational class structure. This expanded middle class needed somewhere to live. The private rented sector was beginning to contract rapidly as a result of local authority clearance and redevelopment and by the sales of better-quality property to owners. The choices were basically threefold: to buy outside London and commute in to work, to buy in the Outer London suburbs and commute in to work, or to rent/buy in Inner London preferably in an area close to the employment opportunities and entertainment facilities of central London. A significant proportion opted to live in Inner London, but this growth in demand could not be accommodated in the traditional middle-class residential areas and prices began to rise rapidly in the first major post-war house price boom of 1970–3.

The solution, perceived by early pioneers in the late 1960s, was to turn to the areas of well-built, aesthetically attractive, centrally located but multiply occupied, decaying, but low-priced areas which had been abandoned by the middle classes and taken over by working-class populations in previous decades. These houses, often large and well laid out, with high ceilings, large rooms and attractive architectural features, offered a large amount of space per pound and the middle classes were very willing to acquire cheap and potentially desirable homes.

On the supply side, private landlords who had been squeezed by years of rent control and low prices were only too willing to take advantage of the new demand and higher prices. As Williams (1976, 1978) has shown, building societies, who were previously very unwilling to lend on inner-city housing, also began to relax their lending criteria. As a result, potential buyers found it far easier to gain access to affordable mortgage finance and the new middle class began to expand residentially into areas which were hitherto predominantly working class. In this way, Chelsea, Fulham, Canonbury, Barnsbury, Primrose Hill and other parts of Islington and Camden began to see the arrival of the new 'gentry'. In the process, the social geography of Inner London has been dramatically changed. David Ley (1981) has suggested that gentrification has thrown the socio-spatial structure of the late-twentieth century city back in time – from the industrial pattern of the nineteenth century and most of the twentieth, which was dominated by an expanded working class, to a post-industrial pattern which is, in many ways, akin to the pre-industrial pattern with the middle classes in the core:

> Present social, economic, and political trends are redefining the morpho-logy inherited from the industrial city . . . If present trends accelerate, the social geography of the nineteenth century city may appear to urban scholars of the future as a temporary interlude to a more historically per-sistent pattern of higher-status segregation adjacent to the downtown core.
>
> (Ley, 1981, p. 145)

In 1961 professionals and managers were concentrated in a few areas of the central boroughs of Camden, Westminster and Kensington, but in each successive decade the distribution spread out into surrounding inner-city areas. This is very clearly seen in Figure 7.3, which shows the percentage change in the proportion of professionals and managers by borough from 1981 to 1991. The boroughs with the highest rates of increase are not the central boroughs but the surrounding boroughs of Tower Hamlets (72 per cent), Wandsworth (63 per cent), Hammersmith and Fulham (57 per cent), Islington (56 per cent), with other large increases in Southwark (46 per cent), Lambeth (43 per cent) and Hackney (41 per cent). When the percentage change in the number of professionals and managers by borough, 1981–91, is plotted against the proportion of professionals and managers by borough in London in 1981 (Figure 7.4), a clear pattern emerges where the greatest percentage increases in the professional and managerial group tended to occur in those boroughs with the lowest percentages in 1981. Although the pattern is not perfect, correlation analysis gives a multiple R of 0.59 and an R2 of 0.34. This indicates the large-scale gentrification of previously working-class areas during the 1980s. A clear divide exists between the Inner London boroughs, with an average increase of 38 per cent in size of the professional and managerial group, and Outer London boroughs with an

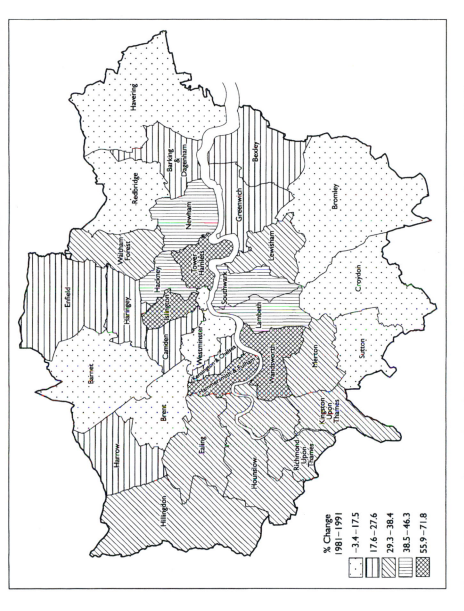

Figure 7.3 Percentage increase in professionals and managers, Greater London, by borough, 1981–91.
Sources: Censuses of Population, 1981 and 1991.

% Change
1981–1991

-3.4 – 17.5
17.6 – 27.6
29.3 – 38.4
38.5 – 46.3
55.9 – 71.8

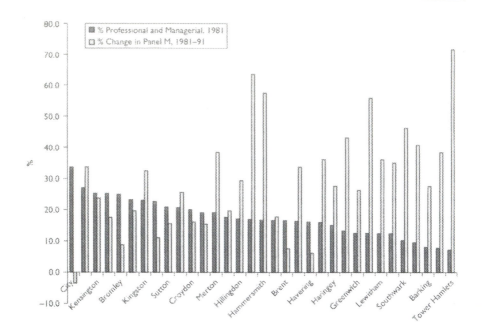

Figure 7.4 Percentage distribution of professionals and managers, by borough, 1981, by the percentage increase in professionals and managers, 1981–91.
Sources: Censuses of Population, 1981 and 1991.

increase of 21 per cent. The East End is still predominantly working class (Figure 7.5) and is likely to remain so for the foreseeable future, but the changes are indisputable. Not surprisingly, the impact has been seen in the owner-occupied housing market where, as Chapter 6 showed, prices have risen far more rapidly in the inner boroughs than in Outer London.

The supply of cheap period houses for conversion to single-family residences in the more attractive parts of the inner city dried up rapidly, aided by the four house price booms of the early and late 1970s, the second half of the 1980s and the second half of the 1990s. As a consequence, potential gentrifiers had to look further afield within London or resort to smaller properties or converted flats. As Chapter 6 pointed out, the phenomenon of converted flats emerged in the 1980s as developers saw the profits to be made by converting houses into flats. By the end of the 1980s conversions were the single largest source of new dwellings in London. More recently, the conversion of old warehouses, factories and offices has pushed the process into city fringe areas.

The sequence of middle-class recolonisation can be traced outwards – north across most of Camden, Islington, western Haringey; east into Hackney,

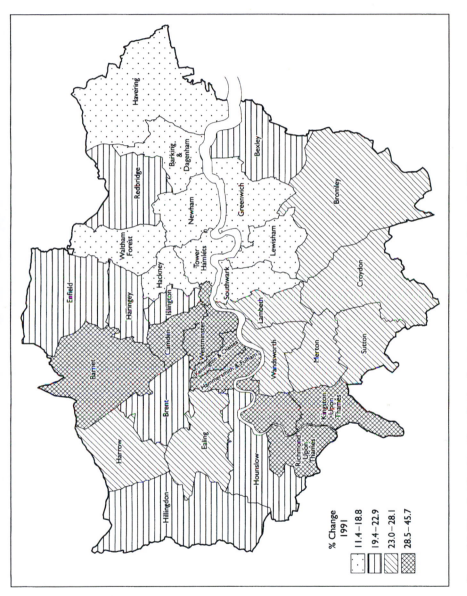

Figure 7.5 The distribution of professionals and managers by London borough, 1991.
Source: Census of Population, 1991.

% Change
1991

11.4–18.8
19.4–22.9
23.0–28.1
28.5–45.7

Stoke Newington and Docklands; west into Hammersmith and Fulham, North Kensington and Ealing; north-west into Kilburn and Brondesbury; south into Vauxhall, Bermondsey, Clapham and the whole river strip; and finally back into the inner areas adjacent to the City of London, Clerkenwell, Shoreditch, Bethnal Green and Aldgate and Whitechapel (Hamnett and Williams, 1980; Munt, 1985; Bridge, 1994; Butler, 1997). As Glancy points out:

> As London property prices shot into the ionosphere at the shiny tail end of the 1980s, those who wanted to live somewhere interesting at low rents shifted ever eastwards. At first, young aesthetes, artists, architects and those who could sense the way the financial wind was blowing took to the shabby yet beautiful streets of Spitalfields. The bars, the restaurants came next, along with booming property prices in that fragment of Shoreditch settled by the trendies.
>
> (Glancy, 1999, p. 13)

To understand the spatial expansion of gentrification and its links to the rise in house prices I would put forward the analogy of a multi-bowl water fountain. The volume of water at the top represents middle-class housing demand. The water falls into the top bowl but, as prices rise, this is soon filled, and the water spills over into the next bowl, which in turn spills over down to the lowest and broadest bowl of the fountain. In this way, gentrification has progressively spilled over into successively wider and cheaper bowls as prices rise and demand spills over. In this interpretation, many gentrifiers in areas such as Stoke Newington and Hackney are there not because these areas were their first choice but because they could not afford to live in Islington. Gentrification can be seen as a form of spatially displaced middle-class housing demand.

The social consequences of gentrification

The social consequences of gentrification are predictable. In a competitive housing market where access is ruled by price, the expansion of the middle classes in Inner London has been associated with the rolling back of the less skilled, the unemployed, the poor and ethnic minorities who have been steadily concentrated into the remaining Inner London local authority estates and the growing housing association sector. Once again, the process was perceptively anticipated by Ruth Glass, who noted that:

> there is very little left of the poorer enclaves of Hampstead and Chelsea: in these boroughs, the upper-middle class take-over was consolidated some time ago. The invasion has since spread to Islington, Paddington, North Kensington – even to the 'shady' parts of Notting Hill – to Battersea, and to several other districts, north and south of the river.

The East End has so far been exempt, although before long some of its districts, too, are likely to be affected. And this is an inevitable development, in view of the demographic, economic and political pressures to which London, and especially central London, has been subjected.

(Glass, 1963, in 1989, p. 138)

Glass went on to point out that:

any district in or near London, however dingy, or unfashionable, is likely to become expensive, and London may quite soon be a city which illustrates the principle of the survival of the fittest: the financially fittest, who can still afford to work and live there.

(Glass, 1963, in 1989, pp. 177–8)

The result of these two processes has, it is suggested, resulted in a decline in social segregation at the macro borough scale in London. There is no longer a small middle-class residential area in central London surrounded by a homogeneous sea of working-class housing. Rather, the middle classes have expanded outwards into a number of hitherto working-class areas. The social class composition of Inner London is now far more mixed than it was thirty years ago. But, at the local or micro level, it is likely that segregation has risen between wealthy homeowners in one street, and unemployed, low-income council tenants a few streets away. Rich and poor, once almost invisible to one another by virtue of residential segregation, are now visible. This has brought greater social tensions. As Glass perceptively pointed out thirty years ago:

the 'colonising' drive of higher classes in London has been accelerated; and so working class quarters are becoming more constricted. Apart from the very rich, it is mainly the young members (or aspiring members) of the middle-upper strata, single people or couples without family responsibilities, who are prepared to pay the exorbitant housing prices of the inner sector. As these people live mainly in one- or two-person households, they have a disproportionately large number of households, and a disproportionately large housing demand, in relation to the total population size of their group . . . There are few hurdles in their path. The working class population of Inner London, old timers or recent recruits, are in every respect 'less equal' not only in terms of income, status, expectations, but also in terms of age and dependency ratios. They have a higher share of old people, young children, large households. They find it more difficult to resist being displaced or hemmed in.

(Glass, 1973, pp. 425–6)

She went on to add that:

traditionally, the most elegant neighbourhoods with the best addresses (such as those in Mayfair) used to be socially mixed (or, as might be said nowadays, polarized). While in such areas the upper class was socially predominant, it was statistically in a minority . . . This kind of mixture is beginning to be much slighter; in such places, the top class (the professional and managerial group) has already become, or is on the verge of becoming, the statistical majority. At the same time, a new kind of apparent class mixture is developing in inner areas which were previously almost wholly working class, and which have now acquired a quite substantial middle-upper class component. But this is likely to be quite often only the first stage of the gentrification process, whereby the working class will be reduced, before long, to the status of a statistical minority in such areas.

(Glass, 1973, p. 425)

Glass was remarkably prescient in her ability to predict the future of social change in Inner London. As she foresaw, gentrification has been accompanied by the departure or squeezing out of the working class from many of its traditional inner-city bastions. By the end of the 1990s, the old East End was under pressure as a growing number of middle-class City workers sought to buy close to work in Whitechapel and Aldgate. In the process, social segregation initially decreases as the middle class move into the working-class areas. Slowly but surely, however, the remnants of the working class are then priced out of the market. If the process continues it is likely that many of the existing working-class city fringe areas such as Bethnal Green and Shoreditch will undergo a process of gentrification. In many ways the East End is now at the cutting edge for gentrification in London. In November 2001 a property column in the *Independent* ran an article on Bow in East London, titled 'East End? Now that's what I call posh' (Liebman, 2001), pointing to property prices for three-bedroom houses of £300,000 and £800,000, to £1 million for large period houses. Prices like these were unimaginable ten years ago, but now the area is seen as convenient to the City and Canary Wharf. There are interesting parallels with the gentrification of New York. First of all, in the 1950s and 1960s Greenwich Village was gentrified; in the 1970s and 1980s Brooklyn Heights, Soho, Tribeca and Chelsea. Then, in the 1990s, it was the turn of the Lower East Side (Smith, 1996; Abu-Lughod, 1994; Smith and deFillipis, 1999). The final challenge is the gentrification of Harlem, which is now under way (Schaffer and Smith, 1986).

Gentrification, social mix and displacement

Although much of the writing on gentrification in the press is consumer orientated and focuses on 'the new hot areas to be colonised', and where gentrifiers can go next, the expansion of gentrification is a two-sided sword, as many writers have pointed out (Orr, 2001). On the one hand, gentrification brings the middle classes back into the inner city, improves housing quality and increases the degree of social mix. In some ways, this is highly beneficial and many urban commentators have for years lamented the social consequences of middle-class flight to the suburbs, the growth of social segregation, the decline of educational standards, etc. On the negative side, the middle-class return to the inner city is necessarily accompanied by either the displacement of existing residents or their replacement by middle-class residents. Either way, property prices and rents rise sharply in affected areas, and existing or potential lower-income residents are effectively priced out of the market. While they may not be directly displaced, they are certainly indirectly replaced. As Orr points out:

> The process was once called gentrification and has always been property led. Inner city areas which had become run-down but which retained spacious period housing attracted middle class families looking for palatial premises at a low cost. As they moved in and refurbished their homes so other like-minded people became attracted to the area, as did the sort of businesses which could service these communities . . .
>
> (Orr, 2001)

Most academic literature focuses on the displacement associated with gentrification and is generally hostile to the process. In housing-market terms, gentrification is taken to have negative rather than positive outcomes for the lower-income groups. This is undoubtedly true, but the problem for many academic commentators is that they want to have their cake and eat it in the sense that they decry the flight of the middle classes to suburbia and their abandonment of the inner city, and they simultaneously criticise the housing market impact of their re-colonisation of the inner city. It is difficult to have it both ways. Either one welcomes the return of the middle classes to the inner city or accepts the class segregation of the past as desirable. It is not possible to reduce class segregation and simultaneously to resist gentrification.

The question of gentrification and displacement has been analysed by a wide variety of authors, principally in an American context. Thus, Marcuse, writing of New York, argues that gentrification and abandonment are two sides of the same coin, in that both are reflections

> of a single long-term process, resulting from the changing economy of the city. This process has two aspects: the shift from manufacturing to services, from reliance on mid-level skills to automation and deskilling, on the one hand, which renders redundant large parts of the workforce and reduces lower-income rent paying ability, and increasing professionalization and concentration of management and technical functions, on

the other, which creates additional higher-income demand for housing. These processes have spatial consequences: blue-collar workers (and potential blue-collar workers) are no longer needed in such numbers downtown; professional and technical workers are in increasing demand there . . . the increasing polarization of the economy is reflected in the increasing polarization of neighbourhoods.

(Marcuse, 1986, pp. 154–5)

Marcuse also argues that in addition to direct displacement it is possible to identify what he terms 'exclusionary displacement'. This occurs when a household is not permitted to move into a dwelling, by a change in conditions affecting the dwelling or its immediate surroundings. He argues it is virtually impossible to distinguish between direct displacement, exclusionary displacement and the pressure of displacement (p. 162).

He is undoubtedly correct regarding the key role of industrial transformation and the professional middle classes in gentrification, but the conditions which prevail in New York and other North American cities are very different from those which prevail in British or European cities. First and foremost, the US has very little social rented housing, whereas in Britain and Europe the proportion is much higher. Second, New York is still characterised by high levels of private renting without the rent control or security of tenure legislation found in Britain. As a consequence, the rental market in New York is far more important and operates much more freely than in London where social rented housing still accounts for a quarter of all households, usually the poorest and most vunerable. Third and very important, British cities have not, until recently, been characterised by abandonment. Where it has happened it is overwhelmingly in older northern industrial cities. The private sector in London has been characterised by housing shortage, not oversupply.

Gentrification and displacement in London

An attempt has recently been made by Atkinson (2000a, 2000b) to measure the extent of gentrification-induced displacement in London using 1981 and 1991 ward level census data. He used ward changes in the proportion of all professionals and managers (both male and female) as a proxy measure for gentrification, and a variety of proxy measures for displacement – namely, a decline in size of the working class, the unskilled, households privately renting, ethnicity defined in terms of those born in the New Commonwealth and Pakistan, the unemployed, the elderly and lone parents. It should be noted that he only took those wards which had a rate of professional and managerial growth higher than the mean rate of increase for London as a whole. This led to the exclusion of all wards with a growth rate of less than 5 percentage points. When these wards were eliminated a total of 133 wards were left out of 755 wards in Greater London. The increase in the remaining wards ranged from 5 to 22 percentage points. Atkinson's results show that the increases in the professional and managerial group were associated with declines in all the proxy displacement variables with the exception of unemployment and lone parents, both of which showed an increase across the board as would be expected (Table 7.1).

Atkinson then produced a regression model which attempts to measure the statistical link between the proxy variables for gentrification and displacement, though, as he readily admits, the model is inverted in that it uses gentrification as the dependent variable and the 'displacement' variables as independent variables. He justifies this on the grounds that 'while gentrification was the motor of displacement logic dictates that displacement would have to take place first in order to vacate dwellings for gentrifiers to subsequently occupy' (2000b, p. 292). The model had a high degree of statistical 'fit': an R2 value of 0.74, and a multiple correlation coefficient of 0.86 (where the range is from 0 [zero correlation] to 1 [total correlation]). In terms of the relative importance of the regression coefficients he found that the decline in the working class was the most important (-0.70), followed by decline in the unskilled (-0.34), and unemployment (-0.17). He concludes that there is 'a high degree of correspondence between the events labelled as gentrification and displacement' (2000b, p. 293).

There is, however, a fundamental problem with Atkinson's analysis which he touches on when he refers to 'the events labelled as gentrification and displacement'. Although he is correct that the increase in numbers of professionals and managers in many wards has been associated with a decline in the size of the working class and unskilled population, in private renting and the elderly, and although he shows that the mean percentage point increase in professionals and managers in the top quartile of wards is associated with a greater proportionate decline in the working class and unskilled and private renting, it is difficult to claim that these changes are necessarily associated with displacement in a direct causal way. Indeed, as he himself admits:

> It was hypothesized that if a negative correlation existed between these changes (i.e. gentrification goes up and the incidence of those termed displacees goes down) some link existed between the two events though such connection cannot be conceptualized as causal or direct.
>
> (Atkinson, 2000b, p. 292)

On the contrary, it is argued that while there is undoubtedly gentrification-induced displacement of the working classes (direct and indirect), the changes in industrial and occupational structure in London over the last forty years have been long-term secular changes associated with the shift from an industrial to a post-industrial economy. In particular, the long-term decline of manufacturing industry and employment has been a direct cause of the decline in the number of skilled and semi-skilled manual workers, just as the growth of financial, legal and business services has been associated with the growth of the professional and managerial group. If we assume that the average working life of an individual is forty years (from age 20 to age 60), the entire labour force will turn over and be replaced during a forty-year period, and a quarter will be replaced every ten years. Consequently, a high proportion of the manufacturing labour force in 1961 will have disappeared by 1991 or 2001. They will have either retired, moved out or died. This change will have taken place largely as a result of long-term industrial and occupational change, not of gentrification *per se*. It is therefore not very surprising that there will have been a marked decline in the working-class population in most wards in London. While this is likely to have been higher in the wards which have experienced rapid increases in the proportion of managers and professionals (and Atkinson's figures show this), by no means all the decline in the size of the working-class population can be linked to gentrification. Indeed, even in the lowest quartile of wards ranked in terms of gentrification, there was a considerable percentage reduction in the size of the working-class population (see Table 7.1).

There is also a fundamental problem with the equation of decline with displacement. While it is undoubtedly correct that a number of working-class residents were directly or indirectly displaced as a result of gentrification, it may also be the case, particularly with some working-class homeowners, that they took advantage of the rise in property values to retire, sell out and move to the suburbs or beyond. To the extent that this is labelled displacement it misleadingly conflates displacement with replacement.

It is also the case that there have been substantial long-term declines in the proportion of private renters in London and in the proportion of the population who are elderly. The decline in the private rented sector began in

Table 7.1 Mean percentage point changes for gentrification and displacement variables, by quartile

Variable	Top quartile	Second quartile	Third quartile	Lowest quartile
Professionals	15.10	11.06	8.93	7.26
Working class	−17.69	−15.84	−14.70	−13.23
Renting	−6.12	−6.12	−3.75	−3.24
Elderly	−3.04	−2.28	−1.71	−1.75
Unskilled	−3.01	−1.82	−1.23	−1.19
Ethnic minority	−0.66	−0.06	0.54	1.19
Lone parent	3.91	2.94	3.96	3.54
Unemployed	12.59	10.88	11.82	11.58

Source: Atkinson (2000a, Table 1).

the 1950s, and, as Chapter 6 showed, it has taken place right across London, both inner and outer, although it has been most marked in Inner London where the proportion of private renters was much higher to begin with. In Inner London the proportion of private renters fell from 64 per cent in 1961 to 28 per cent in 1991, and in Outer London it fell from 30 per cent to 14 per cent. The decline in private renting took place across all boroughs, and it was particularly marked in Inner London in the 1960s and 1970s where large areas of private rented housing were cleared for local authority redevelopment in boroughs such as Islington and Lambeth. The rate of decline in private renting was much less marked in the 1980s, partly because so much had already disappeared, partly because of the Conservatives' sharp cut-backs in subsidies for council house building, and partly because of a recovery in pro-fitability as a result of legislative changes. The point remains, however, that the private rented sector declined across the board during the last forty years as a result of shift in the economics of the housing market from private renting towards ownership, and local authority clearance and redevelopment.

Finally, the decline in the number and proportion of the elderly in London is linked to the changes in demographic structure in the capital. London has a disproportionate number of one- and two-person households, and there has been a substantial increase in the number and proportion of younger people in the capital, particularly as a result of immigration to London both from within Britain and overseas.

Atkinson accepts that it is difficult to causally connect gentrification and displacement using census data, but he nonetheless concludes that 'the strength of the links logically point to an associated and large scale displace-ment process' (2000b, p. 293). This is difficult to accept. While there is clearly displacement the links do not point to large-scale displacement. On the contrary, it is suggested that, at least where the changes in occupational class composition are concerned, what Atkinson's data reflect is the double-edged nature of a process of city-wide industrial, employment and class restructuring. In the transition from an industrial to a post-industrial city, there is an almost inevitable shift from skilled manual to managerial, profes-sional and technical white-collar workers. Atkinson's socio-economic change data measure the decline of one group and its replacement by another grow-ing one rather than displacement. As the sum of the occupational groups necessarily adds to 100 per cent, a decline in one is necessarily reflected in increases in others. It is not solely the growing professional and managerial group who are squeezing manual workers out of gentrifying areas, but the simultaneous decline of the working class in London's transition from an industrial to a post-industrial city (Savitch, 1988; Simmie, 1983; Ley, 1981).

Neil Smith (1996) has argued in his book, *The New Urban Frontier: Gentrification and the Revanchist City*, that gentrification can be inter-preted in terms of the historical revenge of the middle classes for being displaced from the inner cities a century ago. Though Smith's analysis may be metaphorical, this seems unduly melodramatic, although there is little doubt that we are seeing the replacement of the working class and the return of the middle classes to reoccupy houses their forebears relinquished a century or more ago as they fled the inner cities in the face of railway engineering and the large-scale industrialisation.

The future of gentrification

Critics of the importance of gentrification in North America have dismissed it as of limited importance (Bourne, 1993) compared to continuing suburbanisation and inner urban decline, and argue that it is small scale and a temporary product of the post-war baby-boom generation. Berry (1985), for example, described it as 'islands of renewal in seas of decay'. Others such as Badcock (1995) and Wyly and Hammel (1999) take the reverse view and see it as a continuing long-term phenomenon which is reshaping the inner areas of many large cities.

The view taken here is that gentrification will continue to grow in importance in key cities as long as the transformation in occupational class continues. The growth of the professional and managerial middle class, many of whom work in central offices, has generated both a demand for attractive housing and for proximity and accessibility to the centre. Glass (1973) suggested thirty-five years ago that 'the real risk for London is that it will ... become more middle class' (in 1989, p. 178), that gentrification would spread across Inner London and that as house prices rose they would squeeze out low-income groups. Subsequent events have largely borne out her predictions and there is no reason why the situation is going to change. On the contrary, Ley (1981) seems correct to argue that, paralleling the shift from industrial to a post-industrial society, we are also seeing a shift from an industrial to a post-industrial urban residential structure in major cities. It is suggested that current trends will continue and that over the next twenty to thirty years the middle classes will continue to expand outwards across Inner London, aided by the great importance of proximity to work, the attractions of city-centre social life, and the difficulties of long-distance commuting. Whereas nineteenth-century social commentators pointed to the problems of social mix associated with the continued out-migration of the middle classes, today the problem is reversed. The bond dealers, brokers, and creative directors are squeezing out the working class. This process is likely to be intensified as the number of households in London continues to increase and the competition for housing intensifies.

The limits to gentrification?

Four things could limit or halt gentrification. The first is if London loses its place as the leading European financial centre. Though this seems unlikely, it would lead to a significant reduction in the size and purchasing power of the middle class. The second is the expansion of the ethnic minority population of London, and the competition for space which may ensue in some parts of Inner London such as Finsbury Park, northern Notting Hill, Hackney and Brixton. The 2001 census showed that the ethnic minority population of Inner London grew by 48 per cent 1991–2001 and in Outer London by a remarkable 59 per cent. As Peach (1996) has shown, the ethnic minority population of London has spread outward geographically rather than remaining tightly concentrated. Britain does not have ghettos or anything like them, but there are distinct problems in some areas such as parts of Tottenham, Lambeth and Brent, between groups with different norms and styles of behaviour.

The competition is not primarily taking place in the owner-occupied sector except in Outer London where the middle-class British Asian population is growing rapidly, as in Inner London the largely white middle class have won hands down owing to their greater ability to pay and outbid competing groups. Rather, it is taking place between the owner-occupied and what remains of the social rented and privately rented sectors, and the 'battleground' is for education and control of the streets as much as housing.

In London, gentrification has almost completely transformed Notting Hill, Islington, and parts of Hackney where the housing stock consists of larger Victorian terraced houses. But in other, less central and less attractive areas of the city characterised by smaller terraced houses, such as Tottenham, Brent and Clapton, the black population is growing. It could be, therefore, that gentrification will reach its limits in these areas.

The third factor is the growth of violent street crime in London (discussed in Chapter 8). This has been associated with an increased level of killings between rival drug gangs but also of muggings and robberies (Boggan, 2000; Braid, 2000; Sengupta, 2000), and it has generated a considerable increase in fear of crime. Middle-class households have greater freedom of residential choice than other groups, and more may decide to head for the suburbs or beyond, notwithstanding the commuting problems.

The other thing which may limit gentrification is the problem of suitable educational provision for the children of the new middle class. The current education system was not designed with the educational needs of middle-class parents in working-class areas in mind. While the comprehensive system works reasonably well in areas with a good social mix, it can be problematic in poor areas with large working-class populations. As Butler (1999) notes:

> Very few of the elite ever participated in the inner London education market, and until comparatively recently, there were few middle class families with children using inner London's schools. They were dominated by the working class – many of whom have now left. The

middle class has therefore had to reconstruct an education system to meet its need to pass on intergenerational advantage to its children.

(Butler, 1999, p. 90)

Consequently, when children reach the age of 11 and go to secondary school, middle-class gentrifiers face a fundamental dilemma. As Robson and Butler point out:

Among families with school-age children, there is evidence to suggest that education markets are now rivalling those in housing and employment as determinants of the nature, extent and stability of middle-class gentrification of inner city localities . . . [they] experience practical and psychological stresses trying to integrate their ideals of education and of city life with the [perceived] needs of their children and the requirements of successful [middle-class] cultural reproduction . . . Such families almost invariably experience a greater or lesser crisis as children approach secondary school age.

(Robson and Butler, 2001, p. 84)

Intriguingly, reports in June 2002 (Garner, 2002; Owen, 2002) noted that the middle classes are abandoning state schools in London for the private sector. David Hart, the general secretary of the National Association of Head Teachers, stated that:

The fight for the middle classes hangs in the balance. In some areas, particularly in London, they are clamouring to send their children to day schools because of teacher shortages and the impact of bad behaviour. Parents will vote with their feet if they can afford it.

This is not to lament the plight of the middle classes in Inner London. There can be no doubt that, with the financial resources they command, they will always win the battle for private housing. But they cannot win the battle for educational attainment so easily unless the social composition of the inner city shifts rapidly towards the middle class, which is unlikely. Given the high percentage of social housing in most Inner London boroughs, and the collective nature of state educational provision, the middle classes are likely to remain a minority group for many years to come, and in some areas such as Hackney they will almost certainly remain outnumbered. Whether they will throw their weight behind the local state school system, in an attempt to improve provision, or vote with their feet when their children are near secondary school age, remains to be seen.

To some extent, these forces are already in play. There is a substantial out-migration from London of middle-class people in their thirties and forties with young or school-age children (Ford and Champion, 2000). These households are primarily driven by a search for a higher quality of life, better schools and the like. The error made by some observers is assuming that out-migration of these groups must inevitably weaken gentrification. But, as Fielding (1989, 1995) has shown, London, and to some extent the surrounding South East, acts as a social 'escalator region' which attracts large numbers of young people from elsewhere in Britain for higher education, or immediately post-university for a job in the financial or business services public sector or the creative industries. In this way, spatial mobility is linked to upward social mobility and to the maintenance and reproduction of an expanded profes-sional London middle class (Savage, 1988). The in-migration of this group in their early twenties offsets the out-migration of those twenty years older who move out (Hamnett, 1990).

In an intriguing coda to the continuing gentrification of London, the *Independent* ran an article (Akbar, 2001) reporting on Butler and Robson's research. The article, titled 'Middle class settlers fail to join community', was accompanied by a pithy if grossly one-sided editorial, 'Upwardly mobile', which stated that the researchers

> seem stuck in the 1980s, when proletarian romantics resisted all and any change, no matter how welcomed by the working class themselves. Rising house prices? 'Locals' cannot afford to pay them; where will they live? But to whom are rising house prices paid? The social geographers of the UEL are the true inheritors of the coal in the bath tendency. They still live in the days when many people who did not like Mrs Thatcher forgot that all cities change and that London, with its long history of immigration from around Britain and the world, is one of the most dynamic.

The *Independent* is correct that cities change and the influx of middle-class gentrifiers is not necessarily problematic, and may be positively beneficial, but it overlooked the fact that rising house prices also squeeze most locals out of the market. Not many, if any, members of the Inner London working class can afford to pay £300,000 or more for a house. The East End is being gentrified, but few local people can afford such prices.

Chapter 8

Deprivation, social exclusion and its consequences

in Tower Hamlets . . . the council blocks stand facing the commercial sky-scrapers of the City; the one representing one of the most depressed areas in Britain, the other sited in one of the richest square miles in Britain. It is a contrast that has been seen before in London's past.

(GLC, 1985b, p. 8)

The inner city is a social antipodes of middle-class Britain, a universe apart, an alien world devoid of almost every feature of an ideal environment . . . the inner city is now, and is likely to remain, Britain's most dramatic and intractable social problem. Here are concentrated the worst housing, the highest unemployment, the greatest density of poor people, the highest crime rates and, more recently, the most serious threat posed to established law and order since the Second World War.

(Harrison, 1983, p. 21)

Introduction

Earlier chapters of this book have argued that London has not been characterised by growing proletarianisation, pauperisation and immiseration, as some commentators feared in the 1970s. On the contrary, the available evidence points to the growth of a large and prosperous middle class of professional, managerial and technical workers. The growing wealth of this class has enabled them to buy houses and apartments, and to push out residentially into what were hitherto predominantly working-class areas of London. This has been associated with the emergence of fashionable restaurants, bars and shops in what were previously seen as depressed and run-down areas of the city. This is a consequence of London's role as one of the command and control centres of global capitalism. As Friedmann and Wolff put it, such cities 'draw unto themselves the wealth of the world that is ruled by them' (1982, p. 322) and are centres of wealth accumulation and luxury consumption.

It would be a major mistake, however, to infer from this that the problems of London have been solved and that inequalities have reduced or disappeared. On the contrary, London has long been characterised by major inequalities between rich and poor, and these have been intensified during the last twenty years as the earnings and incomes of the top deciles have pulled ahead of other groups. Global cities present a seeming paradox of concentrations of both great wealth and poverty. This has been intensified in recent years by the growth of very highly paid jobs in financial and business services, the decline of manufacturing industry, and the consequent growth of unemployment and economic inactivity. Like other global cities, high levels of inequality in London are a consequence of its global role and its specific industrial, occupational and social structure; they are not an accidental by-product.

The scale and extent of social deprivation in London

Contemporary London may have a much smaller working class than it had forty years ago, and far less absolute poverty, but poverty, deprivation and social exclusion are widespread. The proportion of households with incomes of less than 60 per cent of the median (the standard definition of poverty) has risen sharply in Britain as a whole and within London. This does not mean that such households are now absolutely poorer than they were twenty years ago, but it does mean that the gap between them and the wealthier households has grown. Despite the fact that Inner London is now one of the most prosperous areas in western Europe according to recent EU figures on GDP per capita, Inner London has one of the highest concentrations of deprivation in England.

The Department of the Environment, Transport and the Regions (DETR) 1998 index of local deprivation shows that of the 343 local authorities in England, five Inner London boroughs are in the top 10 most deprived local authorities, and 13 London boroughs are in the top 20 in terms of overall rank (Table 8.1). In terms of the proportion of the population of local authorities who live in wards ranking in the top 10 per cent of deprived wards in England, Inner London boroughs take the top seven places in the ranking and eight of the top 10 places. On the other side of the coin, a number of Outer London boroughs fall near the bottom of the national table of local deprivation.

There is, however, a remarkable paradox in this table, in that Inner London has both the greatest concentration of wards in the top 10 per cent of deprived wards in England yet several Inner London boroughs also rank among the most prosperous areas in Britain, with high concentrations of high-income households and stratospheric house prices to match. Inner London is very different in this respect from many of the older northern or Midlands industrial cities such as Birmingham, Manchester and Liverpool which have high concentrations of deprivation. It is distinctive because of

Table 8.1 London boroughs by deprivation score and overall rank, 1998

	LA score	LA rank	% LA pop. in UK's worst 10% of wards	Ranking (proportion of wards in worst 10%)
Newham	*38.55	2	100.0	2
Hackney	*35.21	4	100.0	1
Tower Hamlets	*34.30	6	99.99	4
Southwark	*33.74	8	88.96	6
Islington	*32.21	10	100.0	3
Greenwich	31.58	11	58.27	16
Lambeth	*31.57	12	89.42	5
Haringey	*31.53	13	79.26	7
Lewisham	*29.44	14	53.58	17
Barking	28.69	15	46.40	22
Camden	*28.23	17	77.72	9
Hammersmith	*28.19	18	65.93	14
Brent	26.95	20	46.86	21
Waltham Forest	26.68	22	66.77	12
Wandsworth	*25.05	30	36.45	35
Ealing	24.48	36	29.59	48
Westminster	*19.05	57	66.29	13
Hounslow	18.89	59	5.07	138
Kensington	*18.54	63	41.82	28
Enfield	16.65	70	14.93	77
Croydon	13.12	88	11.14	91
Redbridge	12.80	90	14.78	78
Hillingdon	8.75	120	0.0	158
Merton	8.31	122	9.33	107
Barnet	7.35	130	9.71	105
Havering	5.52	143	8.13	117
Harrow	5.43	145	0.0	158
Bexley	5.27	148	6.48	127
Richmond	5.00	156	0.0	158
Bromley	3.74	179	12.51	88
City of London	*3.46	183	0.0	158
Kingston	2.4	220	0.0	158
Sutton	0.84	284	5.01	139

Source: DETR (1998), 1998 Index of Local Deprivation.

Note: * Inner London Borough; LA = local authority.

the simultaneous concentrations of wealth and poverty often in quite close juxtaposition, and the use of local authorities as a unit of measurement tends to obscure the local variations within boroughs between desirable and expensive gentrified areas and deprived local authority estates. This paradox of wealth and poverty in sharp juxtaposition is the key to understanding the social geography of Inner London, and the division is generally along the lines of housing tenure as is discussed below. Two-income professional or managerial owner-occupied households can be found living only a few streets away from deprived council estates with high levels of economic inactivity (Vallely, 1999).

Deindustrialisation and unemployment

An important key to understanding poverty and deprivation in modern London is the changing industrial structure. This is as true today as it was in the nineteenth century. London today is a very different city from the nineteenth-century London documented in the novels of Dickens, the illustrations of Gustave Doré, and by the social researchers Henry Mayhew and Charles Booth. Nineteenth-century London was characterised by mass poverty on a scale almost unimaginable today. This was primarily the product of rapid economic change, job loss and increasing casualisation of work which forced a large proportion of London's population into penury from the mid-nineteenth century onwards. In 1850 Mayhew suggested that:

> I believe that we may safely conclude that out of the $4^1/_2$ million people who have to depend on their industry for the livelihood of themselves and families, there is barely sufficient work for the regular employment of half our labourers, so that only $1^1/_2$ million are fully and constantly employed, while $1^1/_2$ million are employed only half their time, and the remaining $1^1/_2$ million wholly unemployed, obtaining a day's work occasionally by the displacement of some of the others.
>
> (Mayhew, 1850, vol. 2, pp. 322–3)

Although Mayhew's figures are open to question, Stedman Jones states in his influential study of poverty in nineteenth-century London, that:

> there is no doubt that the problem he posed was both real and central to the period 1840–1900. Very few workers could expect a working life of stable employment in the nineteenth century, and occupations which appeared relatively immune to the hazards of seasonality, cyclical depression, or technological development – brewery or railway employment for instance – were eagerly sought after despite indifferent wage rates.
>
> (Stedman Jones, 1971, p. 53)

The economy of London saw a dramatic improvement in the twentieth century, aided in large part by the growth of light industry in the capital in the 1920s and 1930s, and migrants flooded into London from more depressed areas of Britain in search of jobs (White, 1986). At its peak in 1961, manufacturing industry employed a third of the total workforce of London but, with the onset of deindustrialisation from the mid-1960s onwards, the period from the 1960s to the mid-1990s was characterised by the continuous decline both of jobs and of manufacturing employment. The number of jobs fell from 4.45 million in 1961 to 3.25 million in 1991 (a loss of 1.2 million) and manufacturing employment fell from 1.45 million to 359,000. Although the decline of manufacturing employment has been common to almost all major cities, the fall in manufacturing employment in London was particularly marked in the 1970s and 1980s and it was not offset by the growth of jobs in the service sector. The consequence was a rise in unemployment (particularly marked among the less skilled) and in economic inactivity rates. Many of the new jobs in financial and business services require high levels of education, which those who had lost manufacturing jobs did not possess. The result was a growing mismatch between skill requirements and local skill levels, which was particularly marked in inner-city areas.

In the early 1980s recession unemployment was lower in London than in the UK as a whole, but in the early 1990s this pattern was reversed with unemployment in London reaching almost 12 per cent in 1993, about 2 percentage points higher than the national average. The big difference, however, is between unemployment rates in Inner and Outer London. In Inner London male unemployment rates measured on the basis of residence exceeded 20 per cent in the mid-1980s and peaked at 25 per cent in 1993, compared to 14 per cent in Outer London (Figure 8.1). The comparable figures for women in 1993 were far lower at 11 and 6 per cent respectively. Although unemployment rates in London have subsequently fallen sharply, the gap between Inner and Outer London remains, and the East End boroughs of Hackney and Tower Hamlets have consistently had unemployment rates some 5 percentage points above the Inner London average (see Figure 8.2). Rix commented that:

> Changes within the industrial and occupational structure . . . have clearly contributed to the dramatic loss of full-time jobs for men, and for women, in inner London East boroughs, growing economic inactivity and rising unemployment levels. The free market interventionist policies and changes in social security legislation have also perpetuated levels of social and labour market exclusion amongst local people. Despite an overall rise in educational performance throughout the London East region the problems of de-skilling, skills mismatch, and the lack of specialised educational qualifications are still widespread.
>
> (Rix, 1997, p. 133)

Buck and Gordon (2000) suggest that a process of 'sedimentation' has taken place within the labour market, which has led to rising levels of structural unemployment among disadvantaged groups who are most residentially concentrated in inner east London. Despite the improvement which has taken place since the mid-1990s, no fewer than twelve London boroughs feature among the worst twenty local authorities in Britain in terms of unemployment. They note that:

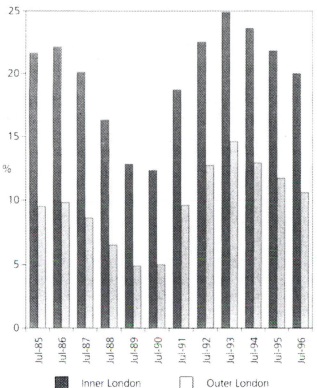

Figure 8.1 Male unemployment in Inner and Outer London, 1985–96.
Source: London Research Centre.

Most of this increasing concentration has occurred within a small number of inner East London boroughs, notably Hackney, Haringey, Newham and Tower Hamlets, where open unemployment rates increased by 6–7 per cent between the 1981 and 1991 Censuses, compared with about 2 per cent in the typical London borough and 1 per cent in the outer metropolitan ring. Increases in the number of 'permanently sick' or on government schemes suggest a further growth of 2–3 per cent or so of concealed unemployment in the first group of areas, against apparent reductions of 1–2 per cent elsewhere in the region.

(Buck and Gordon, 2000, pp. 188–9)

They add that:

Regression analyses at district level suggest that part of this pattern of changes is attributable to more unfavourable trends in labour demand

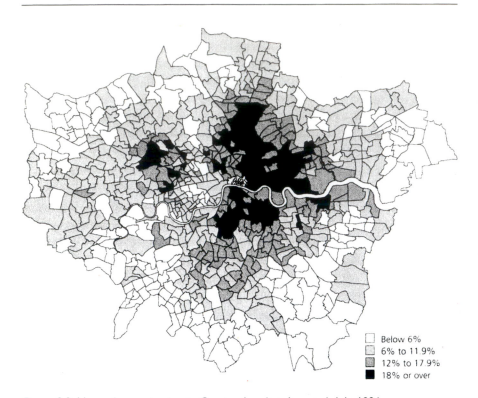

Below 6%
6% to 11.9%
12% to 17.9%
18% or over

Figure 8.2 Unemployment rates in Greater London, by ward, July 1996.
Sources: Department for Education and Employment; London Research Centre.

for semi/unskilled groups in this part of London, and partly to differential growth in disadvantaged sections of the population, notably some of the newer immigrant groups. But together these only account for about a third of the increased concentration of unemployment, most of which seems simply to reflect sharp increases in those places with a history of unemployment, whether associated with population mix, earlier local job losses or a combination of these factors. One important mediating variable in this appears to be the incidence of limiting health conditions, which seems both to reflect past unemployment rates and to significantly boost current ones.

(Buck and Gordon, 2000, p. 188)

One consequence of the rise in unemployment and inactivity rates has been a sharp rise in the number and proportion of recipients of Supplementary Benefit and Income Support (a means-tested benefit paid to those with incomes below a given threshold). Social Security benefits comprise 70 per cent of the income of the poorest fifth of the population, compared to just

Table 8.2 Percentage of households in receipt of Social Security benefits, 1997/8

	Inner London	Outer London	Metro counties	Great Britain
Family Credit or Income Support	25	14	15	15
Housing Benefit	32	16	22	18
Council Tax Benefit	34	21	23	24
Unemployment Benefit/ Jobseekers Allowance	7	5	4	4

Source: Family Resources Survey, Department of Health and Social Security.

11 per cent of the incomes of the rest of the population. The number of recipients rose from 569,000 in Greater London in 1983 to 936,000 in 1994 when the recession of the early 1990s reached its peak, and the proportion from 10 per cent to 17 per cent. In Inner London the rise was more marked: from 15 per cent to 23 per cent. Nearly a quarter of the population of Inner London was dependent on various forms of state benefits to supplement their income at this date (London Research Centre, 1996). Of recipients aged below 60, 12 per cent were classified as disabled, 26 per cent were lone parents and 48 per cent were unemployed. Table 8.2 shows that the proportions of households in receipt of Family Credit or Income Support, Housing Benefit, Council Tax Benefit or Unemployment Benefit/Jobseekers Allowance in 1997/8 are all substantially higher in Inner London than in Outer London, the Metropolitan counties or Great Britain as a whole.

The proportion of adult residents receiving Income Support ranges from 22 per cent in Hackney and Newham, and 21 per cent in Tower Hamlets (the three poorest East End boroughs) to an average of about 8 per cent in most of the Outer London boroughs and 9 per cent in the generally affluent central boroughs of Westminster and Kensington.

The geographical and historical consistency of social deprivation

The geography of unemployment is paralleled by a wide variety of other measures all of which show a consistent pattern. Whether we look at the DETR's deprivation index (Figure 8.3), Jarman's underprivileged area index (Figure 8.4), long-term illness (Figure 8.5), or the distribution of lower quartile household income (Figure 8.6), the broad geographical pattern is much the same. There are marked concentrations of social deprivation in the old East End and inner south London, with a smaller concentration in the inner north-west in Brent. Many of these areas have undergone dramatic physical changes in the last forty years, with large areas of poor-quality

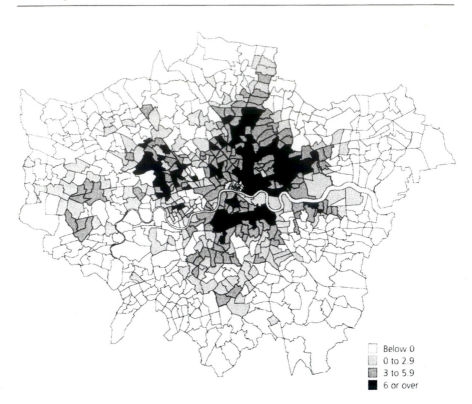

Figure 8.3 Index of deprivation: Z-score, Greater London wards, 1991.
Source: London Research Centre (1996).

rented housing being redeveloped as council estates, and the remaining areas of private housing being gentrified; they are still characterised by high levels of poverty and deprivation. The association of low incomes with high rates of unemployment is scarcely surprising, and both are linked to the distribution of social class and health at both the national and the local level (Congdon, 1995). Indeed, almost all studies of deprivation reveal a consistent pattern of associations with social class and employment (Hamnett, 1976).

It is notable that there is a remarkable historical consistency between the pattern of deprivation in the late 1990s and the distribution of poverty in the nineteenth century which was strongly concentrated in the East End and in inner south London. Green states that although there were pockets of intense poverty within the West End, the inner East End was characterised by massive and widespread destitution:

In 1841 the poorest districts were concentrated to the east and south of the city . . . This poverty belt consisted of districts which contained a

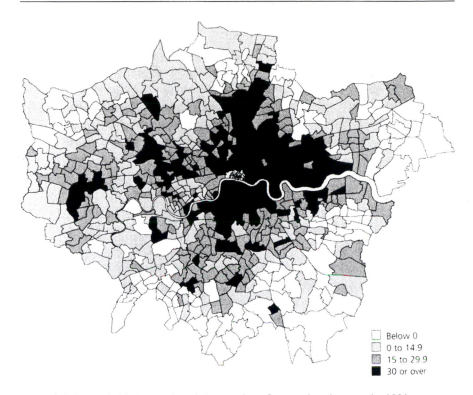

Figure 8.4 Jarman's Underprivileged Area index, Greater London wards, 1991.
Source: London Research Centre (1996).

large proportion of the industrial workforce, including a sizeable pool
of casual labour around the docks.

(Green, 1986, p. 118)

In 1887 Charles Booth published the first of his extensive studies of poverty
in London in the *Journal of the Royal Statistical Society*. This was followed
in 1889 by the first two volumes of his monumental seventeen-volume study,
Life and Labour of the People in London. This comprehensively documented
the scale and the geographical distribution of poverty in the capital on the
basis of detailed street-by-street surveys. He divided the population into
eight groups, from the very poorest to the wealthy, and Green (1986) mapped
the distribution of poverty in 1889 in Inner London. It shows a clear
concentration in the industrial districts around the fringes of the City, in
London south of the river, and in Docklands. This geographical pattern
has persisted today, and Dorling *et al.* (2000) have shown in a statistical
comparison of poverty, deprivation and ill-health in 1889 and 1998 that the

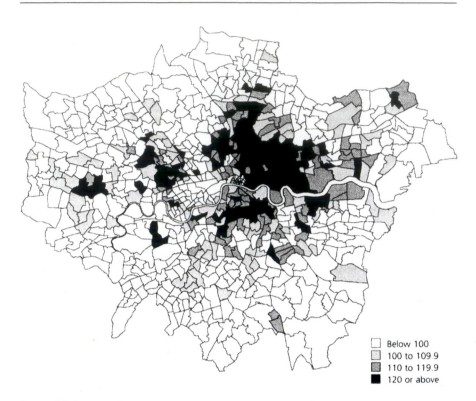

Figure 8.5 Limiting long-term illness, standardised ratios, Greater London wards, 1991.
Sources: Census of Population, 1991; London Research Centre.

Legend:
☐ Below 100
☐ 100 to 109.9
▨ 110 to 119.9
■ 120 or above

distribution of poverty in 1889 is a powerful statistical predictor of ill-health today. They conclude that the distribution of deprivation in Inner London has changed very little over the last hundred years.

Explaining the continuity of social deprivation: class and the housing market

The continuity of social deprivation in particular geographical areas is an important problem. There can be no doubt that the economic roots of deprivation are to be found in the employment, class and income character-istics of areas and their residents. Thus deprivation is concentrated in the areas identified above because these areas have the greatest concentrations of declining employment sectors and poorly educated and less-skilled workers. They are deprived in part because they contain high concentrations of those with the least economic power and the least marketable skills who, faced with a downturn in the demand for their labour, have few other options.

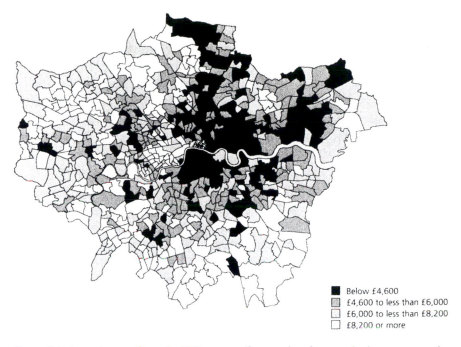

Below £4,600
£4,600 to less than £6,000
£6,000 to less than £8,200
£8,200 or more

Figure 8.6 Annual gross household incomes, Greater London wards, bottom quartile, 1991.
Source: London Research Centre, London Area Transportation Survey.

Indeed, even when demand is buoyant, the structure of the labour market in capitalist economies is such that they command relatively low wages by virtue of their skills. Thus, a great deal of social disadvantage, be it education, housing or health, can be traced back to the class composition of different areas. The higher the proportion of the less skilled or inactive in an area, the lower the labour market opportunities and incomes and the greater the concentration of deprivation.

The importance of social class in influencing life chances has been extremely well documented (Adonis and Pollard, 1997; Sennett, 1972; Savage and Butler, 1996; Halsey, 1988; Reid, 1981), and social class plays an important role in influencing educational attainment, housing and health amongst other outcomes. Chapter 6 pointed to the class composition of housing tenures in London, and class also exerts an important influence on the risk of long-term illness and mortality (Whitehead, 1992). There are also well-documented links between the geography of social class and the geography of health inequalities (Howe, 1986). Congdon (1995) has analysed the relationship between socio-economic status, ill health and mortality in London using data on the socio-economic composition of all of London's 725 electoral wards in 1971 and 1991 and associated health data on mortality and morbidity. He found strong evidence of 'continuing disparities in health and mortality and of a widening gap between socio-economic clusters' (p. 546). The gradients are widest for premature and child mortality but apply to most causes of death, and the disparities in long-term illness were even wider than for mortality. His research found evidence of clear area effects, over and above that indicated by the class composition of areas alone. Curtis (2001) has also examined the geography of health inequalities in London and points to their link with economic conditions. The less skilled and the poorer the population, the greater the incidence of ill health.

But the geographical stability of deprivation also points to housing market continuity as well as class continuity. Indeed, the two are often intimately linked. Many of the poorest areas of contemporary London were poor in the nineteenth century, and were characterised by poor, low-quality housing built for the working classes (Dyos, 1967; Wohl, 1971; Olsen, 1979). In contrast the West End was generally characterised by construction of better-quality housing built for the middle classes, often in the form of closely controlled freehold estates (Jenkins, 1975). Many, though not all, of these middle-class areas retained their social status partly because of the price and quality of the housing they contained. Conversely, areas of poor-quality, working-class housing tended to retain their class character because of the price and quality of their housing.

The principal exception to this generalisation concerns areas of Inner London built for the middle classes in the late eighteenth century and the first half of the nineteenth – such as parts of Islington, Hackney, Camden Town, north Lambeth and North Kensington, which were abandoned to the expanding working classes in favour of newer and more stable middle-class areas. Such areas were converted from single-family occupancy to multi-family occupancy as their class composition changed and by the 1950s they had acquired the characteristics of slum housing. They were overcrowded, poorly equipped with basic facilities and poorly maintained as landlords squeezed the maximum short-turn return out of their tenants. It was not until such areas began to be gentrified from the mid-1960s onwards that the size and class composition of the houses returned to their original form. But, in those areas which were cleared and redeveloped for local authority housing, the class character of the areas has tended to remain firmly fixed by virtue of the design and tenure of the housing. Only in a small number of cases, where local authorities have sold blocks to developers who have remodelled them for private housing or where the location of the area is such as to attract middle-class interest, has their class composition changed dramatically. To a significant extent, the physical character, and the price and tenure structure of the housing market, tends to reproduce the class character of different areas. As Robson (1975, p. 6) commented in a deceptively simple observation, 'social areas are comprised of people, living in houses, distributed in space'. Change the type, quality or distribution of housing and the social character of the area is likely to change. Conversely, if the population of a poor inner-city area is rehoused in new estates, the class and social composition of the area is likely to be directly reproduced in the new housing area. Hence Olsen's (1979) comment that 'In apportioning praise and blame for the condition of life in Victorian London we must remember that builders built buildings: they did not build poverty, inequality, disease, injustice, ignorance and other social evils' (p. 266). Arguably, however, they did ensure its reproduction in some places rather than others.

In a housing market where residential choice is dominated by ability to pay, low skill, unemployment or economic inactivity will almost inevitably result in low income and social renting and relative immobility. Thus, if the local economy experiences a sharp downturn, as happened with Inner London manufacturing in the 1970s and 1980s (Rix, 1997), most local residents are unable to move to search for jobs. They are the victims of what the Lambeth Inner Area Study (DoE, 1976) termed 'the housing trap'. It is useful in this context to point to the correlation between the distribution of unemployment, low incomes and council housing (Fieldhouse, 1999). This is not to suggest that council housing in any sense causes disadvantage, but that the disadvantaged are more likely to live in council or social rented housing. In addition, a concentration of the disadvantaged in poor council housing may generate an 'area effect' whereby the cumulative effects of concentration produce additional problems and disadvantages. This is likely to be most marked in the least desirable and most problematic estates. There are a large number of such estates in Inner London, typically characterised by high-density, or high-rise, system-built concrete-slab construction in the 1960s or 1970s, though some estates date from the inter-war period or earlier. As Chapter 6 argued, such estates tend to be at the bottom of the pile in terms of desirability and have tended to concentrate individuals and households with the least market power and knowledge of the allocation system and the greatest immediate need of housing. As a consequence, they contain high concentrations of the economically inactive, long-term unemployed, certain ethnic minorities and female lone parents, almost all of whom are on low incomes and have restricted housing choices. The concentration of the economically inactive in the social rented sector was shown in Chapter 6, and of some ethnic minorities in Chapter 5. Similar concentrations are found of lone-parent families and low incomes. With the out-migration of skilled and white-collar workers during the 1960s and 1970s, council housing in the inner cities has increasingly become home to those with low skills, low incomes and limited housing choice. As such, it concentrates the most deprived into a sector which is steadily declining, and where many of the better suburban homes have been sold to right-to-buy tenants. As Harrison put it in his penetrating study of social deprivation in the Inner London borough of Hackney:

The ideals of council housing that motivated Hackney's working class councillors in the first three post war decades were noble: to take families out of those overcrowded tenements, with shared or absent baths and toilets; to remove them from exploitation and insecurity; to provide decent, sound houses, each with their own bath, toilet and kitchen; to give tenants a humane and benevolent landlord without an interest in profit.

He adds that:

There may have been a time, in the 1950s perhaps, when these dreams were roughly approximated by reality [but] in the inner cities of the early 1980s they ring hollow. Many of yesterday's pristine estates have become today's slums, some of them fit only for demolition, many more requiring massive expensive repairs and remodelling to make them anywhere near habitable . . . Social ghettos in terraced Victorian streets have been reproduced, in more destructive form, in Elizabethan blocks. Private squalor in rented rooms has been replaced by public squalor in rented flats. The state of public housing . . . has become one of the central problems of the inner city.

(Harrison, 1983, pp. 203–4)

Harrison suggests that for a large proportion of Hackney's residents, their 'principal ambition is to get out of Hackney'. These problems are not just confined to Hackney; they are found in almost every Inner London borough. One resident of the Ethelred estate in Lambeth stated that:

'I don't like the Ethelred – it's a dump . . . the Ethelred has some serious problems . . . in terms of the structure of the estate. I can't help but feel the social side is a reflection of the structural side – the design and outlook and density of it all . . . that's probably the pitfall of the area – it's too dense . . . The roads don't look like roads, they look like alleyways . . . They've got these concrete slabs all over the show – and I think it makes it look like rabbit warrens.'

(quoted in Beaumont and Hamnett, 2002, p. 31)

Social segregation and its consequences

Social segregation in cities is not new. Rich and poor have been geographically divided since the start of the Industrial Revolution. The advent of mass urbanisation and industrialisation during the nineteenth century saw the creation of working-class slum areas on a scale hitherto unknown. This was noted in Manchester and the other northern industrial cities by Engels in the mid-nineteenth century (1969), but was particularly marked in London where the growth of population from 1 million in 1801 to 6.5 million in 1901 was accompanied by the rise of a new urban working class concentrated in the East End. Why should social segregation be regarded as a potential problem? The conventional right-wing view is that it is not a problem, that 'like chose to live with like', and that social segregation minimises social friction between different groups. But, while this may be convenient from a Conservative perspective – that is, the problems of the poor can be ignored as long as they live elsewhere and do not intrude on middle-class prosperity – it is both selfish and short-sighted, to put it mildly. If we look at segregation from the viewpoint of the distribution of social opportunity, it is clear that concentration of the less skilled, poor and disadvantaged in the worse areas reinforces social exclusion, limits opportunities for housing, work, education and social mobility. It can also lead to adverse consequences in terms of the creation both of attitudes and of behaviours and a culture of low or weakened expectations which work against social mobility.

There is also a second argument that excessive concentrations of the less skilled and the poor constitute a threat to the existing (middle-class) social order. This view was most marked in the late nineteenth century, when the 'residuum' seemed to pose a threat to the overall social and political stability of the capital. Octavia Hill, the social reformer, pointed out the perceived dangers of reproducing working-class segregation in the East End:

> I feel very differently about districts like Marylebone and districts like Whitechapel; where you get acres and acres of poor people together, it is a very good thing to move some of them; but where you have a large body of richer and more educated people living in the neighbourhood, and there is a great demand for labour, it is important to keep the poor amongst them; if we send people to the suburbs, they go to cheaper suburbs, and we create for ourselves over again such districts as we are lamenting to find in the East End of London.
>
> (SC 1882 (housing) q.3, 248)

Stedman Jones points out that by the 1890s, Octavia Hill's prediction had come true as a result of the absence of facilities for working-class railway travel. The LCC noted in 1898 that the Great Eastern was the only company to provide for this. As a result:

> While there is an enormous working class population in that one direction, there are on the north, west and south of London, within a much less distance of the Bank of England, large tracts of land not developed for the erection of houses: sometimes even within the county boundaries. *The aggregation of so vast a population of one class in one locality in this way seems to be productive of social danger.*
>
> (quoted in Stedman Jones, 1971, p. 323; emphasis added)

The modern argument for greater social mix is essentially founded on the notion that greater mix provides greater opportunities for the disadvantaged, increases the overall social benefit and reduces social costs by virtue of the fact that more of the population are integrated rather than excluded. Essentially, the argument for greater social mix is that it maximises social opportunities for the greatest number of people (Sarkisson, 1976).

Social exclusion

The concentration of social problems in the poorest neighbourhoods was recognised in the government's Social Exclusion Unit (1998) report, *Bringing Britain Together: a National Strategy for Neighbourhood Renewal*, which stated in the preface that:

> Over the last generation, this has become a more divided country. While most areas have benefited from rising living standards, the poorest neighbourhoods have tended to become more rundown, more prone to crime, and more cutoff from the labour market. The national picture conceals pockets of intense deprivation where the problems of unemployment and crime are acute and hopelessly tangled up with poor health, housing and education. They have become no go areas for some and no exit zones for others.
>
> (SEU, 1998)

In his introduction to the report, Prime Minister Tony Blair stated:

> over the last two decades the gap between the 'worst estates' and the rest of the country has grown. It has left us with a situation that no civilised society should tolerate. It is simply not acceptable that so many children go to school hungry, or not at all, that so many teenagers grow up with no real prospect of a job and that so many pensioners are afraid to go out of their homes. It shames us as a nation, it wastes lives and we all have to pay the costs of dependency and social division.

Although, as Watt and Jacobs (2000) have pointed out, there is a danger that council estates may be demonised as the urban 'other' (the late twentieth-century equivalent of the East End), the issues addressed in the report are very real, and it showed that the concentration of some problems, such as unemployment, has increased over the last twenty-five years in some of the worst estates. There are a number of dimensions to the issue of social exclusion, including unemployment, the incidence of crime, lack of mobility, poverty, and low education. In her national study of disadvantaged areas Lupton (2001) notes that most wards with high concentration of poverty in 1991 were characterised by a high proportion of working-age population not working, studying or on a government training scheme (double the average for all other wards), levels of unemployment triple that of other wards (24 per cent against 8 per cent), households without a car (triple other wards). She also noted that most 'poverty' wards had a very high proportion of public housing (mean 53 per cent compared to 16 per cent for all wards). Nearly two-thirds (63 per cent) of poverty wards had 50 per cent or more households in public housing.

Education

One of the clearest examples of the impact of geographical variations in social class is seen in educational attainment. It has long been known that educational attainment is class related (Jackson and Marsden, 1966), and that this phenomenon has a strong geographical dimension. Areas and schools with high proportions of the lower social classes generally have lower educational attainment than areas and schools with high proportions of the higher social classes. This is not just a result of differences at the individual or household level. There are strong 'area effects', and areas and schools with high proportions of disadvantaged groups tend to perform worse than would be expected because of specific problems such as the high proportion of children with English as a first language and the development of shared attitudes and expectations which are antithetical to the value of education, as Robson (1969) showed in his early work on educational attainment and social class in Sunderland. In his provocative and penetrating book on social deprivation in the London borough of Hackney, Harrison (1983, p. 263) described the school system as being one of 'Schooling for Failure'. He argued that: 'After housing and the local economy, education is the third of the key links in the chain that binds the inner city and its children to their destiny'. He pointed to two aspects of family background which affect schooling. These were, first, the effects of poverty, overcrowding and over-stretched parents on family inputs into education, and, second, the impact of family conflict, instability, disciplinary problems, particularly for single parents, and the attractions of truancy, delinquency and street life on the pupil's behaviour.

In a wide-ranging review of recent research on educational attainment Sparkes (1999) points to the strong associations between the social characteristics of catchment areas and their pupils and low levels of educational attainment. Thomas and Mortimore (1996) found that between 70 and 75 per cent of school variation in attainment is explained by pupil intake factors. The key factors are pupils' personal characteristics, such as prior attainment, gender and health; socio-economic factors including low income, parental unemployment, social class and housing; parental educational attainment; family structure, including family size, lone parent status and institutional care; ethnicity, including ethnic group language and fluency in English, and other factors including parental motivation and interest. There is a high level of interdependence between these factors, and low income, as indicated by free school meal eligibility, is strongly correlated with low levels of educational attainment at all levels (West *et al.*, 1999). Variations in this indicator account for some two-thirds of the variance in attainment between local authorities, and this correlation has been replicated at the school level.

It is thus not surprising, given the marked differences in class composition between the London boroughs, that there are marked variations in educational attainment. These differences have been intensified in recent years by changes in ethnic composition and a growing proportion of school children who do not have English as a first language. The standard measure of educational attainment is the proportion of children achieving 5 or more GCSE passes at grades A to C in examinations taken at age 15. The average for England in 1999 was 48 per cent. The range of averages for state schools, excluding the independent or private fee-paying schools, in the Inner London boroughs ranged from 49 per cent in the relatively prosperous borough of Hammersmith and 47 per cent in Camden, to 27 per cent in Islington and Hackney. The range of scores in Outer London is very different: they run from 34 per cent in working-class Barking to 60 per cent in middle-class Sutton, with an average score for Outer London of 50 per cent compared to an average score for Inner London boroughs of 35 per cent (Table 8.3).

These differences in educational attainment at age 15 have considerable importance because they have a major impact in terms of the chance to go

Table 8.3 Average percentage of pupils achieving GCSE grades A–C, London boroughs, 1999

Inner London	Average	Outer London	Average
Hammersmith	48.9	Sutton	60.1
Camden	47.1	Redbridge	58.5
Kensington	44.5	Kingston	57.0
Wandsworth	35.6	Barnet	56.9
Westminster	35.0	Bromley	56.5
Newham	34.6	Harrow	55.1
Lewisham	32.3	Havering	54.2
Lambeth	31.9	Richmond	51.9
Greenwich	31.7	Bexley	47.9
Tower Hamlets	31.1	Enfield	46.8
Haringey	30.3	Hounslow	46.7
Southwark	29.5	Ealing	45.6
Hackney	27.8	Hillingdon	44.1
Islington	27.4	Brent	43.0
Waltham Forest	36.6	Croydon	42.4
Barking	34.0	Merton	38.8
Mean score	34.9	Mean score	50.3

Source: Department for Education and Science (1999).

on to higher education for a further two years and then go into higher education. Pupils who get less than 5 GCSE grades A–C are very unlikely to be accepted to go into the sixth form to study for A levels and are thus most unlikely to enter into any form of higher education. These educational credentials also affect employability and likely future earnings. Gordon (1996) points out that 'individuals without formal education or craft qualifications face at least double the risk of unemployment' and with the decline of skilled manual jobs in London 'educational qualifications have become the only route to reasonable job security' (p. 408). Gordon also suggests that:

> In London, the extreme case of de-industrialisation, a polarisation of potential work 'careers' has been noted such that those (particularly males) without qualifications are increasingly reliant on finding jobs in the unstable, competitive sectors of services, where the risk of recurrent unemployment is particularly high. And whereas in the more secure office services a high proportion of the qualified workers will be drawn from the other parts of the country, with graduates being particularly mobile, the bottom end of the labour market is much more likely to be indigenous. Hence the question of whether local children achieve basic levels of education . . . has a more direct significance for the structure of the local labour force than does the level of achievement among high flyers.
> (Gordon, 1996, p. 408)

Gordon (2000) notes that Greater London presents a paradox of 'having one of the best qualified labour forces in the UK, but some of the worst school examination results' (p. 18). Thirty per cent of London residents have degrees or professional qualifications, against 23 per cent nationally, which confirms Fielding's (1991) analysis that London functions as an escalator region, sucking in university students and graduates from elsewhere in Britain or overseas. London has a very professionalised workforce, but it does not make it any easier for London residents to get jobs if they lack relevant educational qualifications. On the contrary, labour force requirements in London may be helping to create a group of long-term unemployables.

Crime and deprivation

One of the social consequences of social deprivation is an increased incidence of and exposure to crime. Home Office data show that the number of notifiable offences recorded by the police in London rose from just over 8,000 per 100,000 population in 1981 to just over 12,000 in 1993. But the number of violent crimes recorded by the police more than doubled over the same period: from 400 per 100,000 population to 950. The increase was 134 per cent, which was the second highest of all regions. Not surprisingly, the boroughs with the highest levels of notified violent crimes in 1993/4 were in

Inner London, and the LRC found a correlation between the number of violent crimes and the index of local conditions. The most deprived boroughs generally had the highest levels of violent crime (London Research Centre, 1996)

There has recently been major debate in the media regarding the rise in violent crime, and one worrying development in recent years has been the sharp rise in the number of gun-related crimes and killings in Inner London, particularly in areas with a high proportion of ethnic minorities such as Brent, Lambeth and Hackney where carrying a gun or a knife is now seemingly commonplace among young males. The problems are particularly severe in some of the most deprived council estates where many residents are afraid to venture out after dark for fear of attack by one of the gangs which roam estates (Sengupta, 2000; Boggan, 2000; Braid, 2000; Cobain, 2000). Ethnic minority groups are most at risk and the 1992 British Crime Survey found that the percentage of respondents victimised in terms of personal offences ranged from 9.7 per cent of whites to 13 per cent of Asians and 14 per cent of Afro-Caribbeans. These concerns were crystallised by the killing of Damilola Taylor, a young Nigerian boy in a deprived council estate in Peckham in 2001. This case, and that of Kayser Osman, a 15-year-old Somalian refugee who was stabbed in a deprived inner-city council estate in Brent (Milmo, 2000), has generated massive press coverage about the breakdown of social order and the rise of uncontrolled children with no social values. These incidents have been well publicised, but they are not isolated cases. There has been a sharp rise in the incidence of violent street crime and the government have recently announced their intention to crack down on it.

Mingione (1996) has suggested that a general concern about the 'new urban poverty' in its various guises is due to 'a barely concealed nightmarish vision in which every large city is becoming Americanized and divided up into clearly separate areas of great affluence and "ghettos" of chronic poverty and marginalization, the site of riots and generalized violence' (p. 324). For many residents, this vision has become a reality – it cannot be dismissed as a 'moral panic' constructed by the mass media. A resident of the Clapham Park Estate in Lambeth, interviewed as part of the EU social exclusion project URBEX, commented that:

> 'Six years ago you suddenly had an increase and influx of crime, almost overnight. There was the clearing up of Streatham. Pimps, prostitutes, their dealers, the whole lot moved "lock, stock and barrel". Even early in the morning you began to see "working women" trying to stop cars on the main road. They had a serious drug problem – every so often you'd see them crying and shaking . . . a sense of fear began to develop. The crack dealers had moved in. Mugging began to increase.'
>
> (quoted in Beaumont and Hamnett, 2002, p. 43)

Another resident, a Portuguese woman, commented that:

> 'I think this is a bad area. The drug dealers all the time. One drug dealer goes and another moves in . . . It's very scary. Having children is very difficult – they like to play outside. It's not very safe. In the day and at midnight it's very bad. When you want to walk down the stairs you just see all these drug dealers smoking in the stairs. They see you but you have to say "excuse me" to get past. It's very scary because they look at you like they want your money – dirty look . . . People go over there to buy drugs and then the prostitutes . . . We hear them in the night, shouting and swearing . . . All problems are getting worse.'
>
> (quoted in Beaumont and Hamnett, 2002, p. 62)

Another resident confirmed this:

> '[The neighbourhood is] definitely worsening. The more and more they heat up the war against the dealers in the Town Centres the more we get them. In the last four years suddenly we've become the "Crack Park". Drug fallout is the key. In the 80s the worse threat of violence would be getting beaten up, now you could be stabbed or shot. People are scared with the young people around. We've been lucky that there haven't been any bad gangs, but they're starting to emerge.'
>
> (Beaumont and Hamnett, 2002, p. 63)

One consequence of the rise in crime is an increase in the number of private 'gated estates', and there are plans in at least four local authorities in London to create private police forces to patrol the streets and deal with anti-social behaviour. In April 2002, Ian Blair, the Deputy Commissioner of the Metropolitan Police, announced details of a scheme to recruit 300 civilian guards to patrol the streets in a 'second tier of policing'. He warned that London is in danger of being divided between rich and poor with the rise of American-style security ghettos and what he termed 'justice by geography'. In the wealthy borough of Kensington and Chelsea, the council plans to set up its own force of 120 'borough constables' at a cost of £5 million a year. Merrick Cockrell, the leader of the council, stated that 'Residents are concerned about crime and their safety in their homes and on the streets. The reality is that there are not going to be enough police officers on the streets to satisfy public demand' (Bennetto, 2002).

Implications

Inner London has had major concentrations of social deprivation for over one hundred years. Although many of the worst areas of housing have been redeveloped as council estates, and there have been a series of government policy initiatives aiming to reduce urban deprivation since the mid-1970s, very little has changed. Unemployment is still above the London average, as is the incidence of poverty and supplementary benefits. Educational attainment is still far below that of Outer London and the incidence of ill health is higher. The social problems described and analysed by Paul Harrison (1983) in Hackney and other areas still exist. What has changed is that the gradual expansion of the middle classes into gentrified areas of Inner London is slowly pushing back the geographical extent of these problems into a series of more concentrated areas of poor-quality social rented housing. In the process, the gap between the large well-educated and well-paid middle classes, and the less skilled, the unemployed and economically inactive is arguably increasing. Forty years ago, the less educated and unskilled were generally employed in unskilled jobs, but these have slowly disappeared. The result is the growth of a poorly educated, low-skilled and increasingly unemployed workforce. Although London is characterised by a high proportion of both highly skilled jobs and highly educated workers, many of these workers migrate or commute in from outside London, and it has a substantial proportion of poorly educated residents for whom the changes in the structure of the London labour market have not been beneficial. While proponents of polarisation argue that labour market trends in global cities are creating a two-tier job market, the reality seems to be the disappearance of many less-skilled jobs, including large numbers of public sector transport jobs such as railway porters, bus conductors, guards and the like. The jobs that remain in this sector, and those in the cleaning and security areas, are often taken by either new unskilled migrants or ethnic minority groups with little option but to work at the minimum wage. Wages of £4 an hour, or £160 for a forty-hour week, do not go very far in London where weekly travel costs can take £20 of after-tax income. Not surprisingly, the gap between those earning £160 a week or those dependent on state benefits, and those earning £1,000 or more a week, has continued to grow. Growing earnings and income inequality are one of the hallmarks of global cities like London. While high-skilled jobs and incomes are beneficial, they place increasing pressure on the housing market for those without the skills to gain a foothold in the growth sectors.

Chapter 9

Remaking the landscape
From industrial to post-industrial city

The 'global cities' that have captured attention in recent years owe much to intense competition in international financial services. Partly as a matter of providing world class commercial facilities, and partly as a matter of image-creation, the effort to attract geographically mobile investment activity changes a city's perspective. The old, diversified urban centre is cleaned up for new offices and cultural consumption; in the process, it becomes more expensive. Not surprisingly, government priorities shift from public goods to private development.

(Zukin, 1992, p. 195)

Post-industrialism can be seen as a transformation of the built environment: factories are dismantled, wharves and warehouses are abandoned, and working class neighbourhoods disappear. Sometimes there is replacement of one physical form by another – the growth of office towers and luxury high rises or refurbishing of old waterfronts. Cafés and boutiques arise to feed and clothe the new classes.

(Savitch, 1988, p. 5)

Introduction

One of the key arguments of this book has been that London, like other global cities, has experienced a shift from an industrial to a post-industrial economy since the mid-1960s. This has had major consequences for the nature of work and the structure of occupations as manual jobs in manufacturing have been replaced by non-manual jobs in financial and business services and the cultural industries. This shift has also been associated with a change in production requirements and type of workplace. Factories are no longer needed, whereas modern office space is (Coupland, 1992a). In addition, changes in transport and communications technology have rendered rail goods yards, canals, wharves and docks redundant. Large parts of the older physical infrastructure of cities which were crucial for production and distribution in the industrial era are no longer needed in the post-industrial era and have fallen into decay and dereliction (Savitch, 1988). Conversely, the growth of London's role as a financial and business centre has seen the rapid expansion of the City of London and the emergence of new 'landscapes of financial power', such as Canary Wharf (Zukin, 1992), as finance capital has overtaken industrial capital in importance and the requirements for space and built form have changed. As Ball (1994) points out: 'Buildings are the spaces within which much of all economic activity takes place so alterations in those basic economic characteristics fundamentally influence the demand for buildings' (p. 672).

Changes in built structure and urban form do not simply happen of their own accord: they are the result of specific decisions driven by specific processes, and it is generally agreed by academics (Beauregard, 1991; Fainstein, 1990, 1994b; Harvey, 1978, 1985; Zukin, 1992) that the restructuring of capital is crucial to understanding the process of urban change. As Beauregard put it:

> As capital restructures, it also restructures space. Places of limited profitability are devalued, even while new economic activities transform peripheral areas into sites of massive capital investment. The restructuring of the labour process subsequently reaches into residential areas and commercial activities to express spatially the inequalities of income and wealth derived from employment (or the lack of it) and investment (or disinvestment).
>
> (Beauregard, 1991, p. 91)

Beauregard (1991) argues further that the economic restructuring of global cities has 'resulted in *three* significant transformations of the built environment of core cities' (p. 91). These are, first, the establishment of international producer services in high-rise office towers of central areas; second, abandonment of large-scale factories and expansion of small-scale and mixed-use industrial districts which accompanied downgrading of the manufacturing sector; and third a rise of immigrant and gentrified neighbourhoods. Beauregard and others (Soja *et al.*, 1982) point to the importance of the recession in the early 1970s following the first oil price rise. This period of economic crisis and the attendant stagflation, rising energy costs and plant closures marks a sharp break with the previous period and there is a parallelism between economic restructuring and urban restructuring, as happened in earlier periods of economic restructuring in the late nineteenth and early twentieth centuries. This is precisely the argument made by Stedman Jones (1971) in his analysis of the linkages between the economic crises of late nineteenth-century London, the expansion of the City, the growth of casual labour and pauperisation, and the resultant housing crisis. Although the sequence of post-war urban land-use changes outlined by American analysts differs considerably from those in Britain, not least because of massive suburbanisation and inner-area abandonment which took place in many American cities post-war, it is clear that there are strong links between economic restructuring and urban land-use changes in both countries. Chapter 2 outlined the dramatic changes which have taken place in the industrial structure of London, with the shrinkage of manufacturing industry and employment and a growth of the financial and business services sector and the creative industries. Not surprisingly, these dramatic shifts in industrial and employment structure have been reflected in the built environment, and Fainstein (1990) argues that the changes in both London and New York reflect the growing dominance of economic factors in urban development. Local government planning policy has increasingly aligned itself with the needs of capital accumulation for land-use transformation in global cities. In many respects, it has had no choice. Local government cannot stop industries closing down or relocating abroad and they need to respond to the requirements of finance and business services for more office space. As Fainstein puts it:

> During the 1980s two global cities, New York and London, have shown a remarkable convergence in public policy and development patterns. Where previously they had followed quite different planning traditions, their governing regimes have responded similarly to world economic forces that have heightened the importance of financial control centres while diminishing the significance of manufacturing locations.
>
> (Fainstein, 1990, p. 553)

This said, changes in planning policy are not pre-ordained or determined by the needs of capital or development, and central and local government planning can respond in a variety of ways to the situations they face. Nowhere is this more true than in the case of London's Docklands where, faced with the problem of massive decline and decay, central government initiated a policy of large-scale redevelopment under the auspices of the London Docklands Development Corporation. In the event, this has proved to be a spectacular commercial success, though in the depths of recession in the early 1990s it looked as though it could prove to be an expensive failure. The financial collapse of the Canary Wharf scheme in the early 1990s, and its subsequent recovery, highlights the importance of cycles of boom and bust in development (Ball, 1994; Barras, 1983; Henneberry, 1999). But, before considering this and other recent developments, it is necessary to look at the impact of the virtual disappearance of manufacturing industry on the landscape of London.

The demise of manufacturing industry in London

The impact of this change has been profound. In 1964 Peter Hall described what he termed as Inner London's 'Victorian Manufacturing Belt' which ran around the City:

> It is dominated by the great industrial crescent which runs around
> the north and east sides of central London, from the western edge of
> the City and the West End, through the southern parts of St Maryle-
> bone and St Pancras, through Islington, Finsbury, Holborn, Shoreditch,
> Bethnal Green and Stepney. It extends to Southwark on the south
> bank of the river; to the north it throws out two great projections, one
> north and north west to Kentish Town and Holloway, one north-east
> to south Hackney and Stoke Newington . . . this great crescent lies
> wholly within that area of London which was fully built up before
> 1900.
>
> (Hall, 1964, pp. 226–7)

Most of the old Victorian manufacturing belt has now disappeared. The
same is true of many of the larger manufacturing plants spread across Inner
London and the port processing industries. The last major industrial employer
in London is now Fords at Dagenham, and they are shortly to cease car
production and switch to engines-only production. The visual legacy of the
disappearance of manufacturing industry is seen in the number of derelict
factories and workshops in Inner London. Many have been transformed
either into retail parks (for example the transformation of the old Hoover
factory on the A40) or, depending on location, into desirable loft apartments
(for example, the Bow Quarter which used to be Bryant and May's match
factory). The last large industrial employer in Hackney (Lesney toys) closed
down when production was shifted to South East Asia. In the old city fringe
(Figure 9.1) of Shoreditch, Clerkenwell, Hoxton, Whitechapel, and parts
of Southwark, the printing industries, the furniture, engineering and food
manufacturers have all gone. The former Allied Brewery site in Clerkenwell has closed down and is being redeveloped for offices and luxury apartments, along with the old Danish Bacon factory just north of Smithfield. The
former Truman Brewery in Brick Lane just to the east of the City of London,
which closed in the 1980s, is now home to a large internet exchange centre
(Joseph, 2000), and many other industrial buildings in the city fringe are in
the process of being converted to luxury residential or other post-industrial
uses.

  There are two sides to this transformation. On the one hand changes
in the economic and spatial organisation of production (such as takeovers
and closures in the brewing industry and the concentration of production of
a small number of large sites, or shifts in food production outside London)
have rendered many traditional inner-city industrial land uses redundant.
On the other hand, the growth of financial and business service jobs in
London, the need for additional office space, and the desire of a proportion
of the much expanded post-industrial workforce to live close to work and
entertainment facilities have provided developers with a profitable market
for land-use conversions.

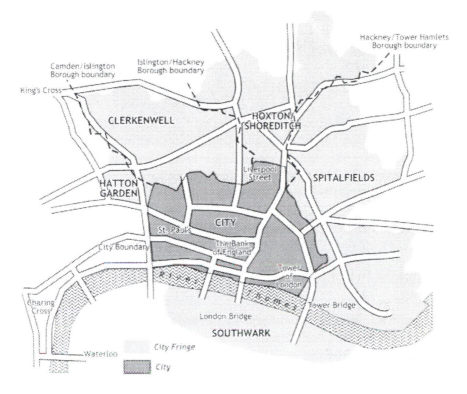

Figure 9.1 The City Fringe.
Source: The City Fringe Partnership.

In addition to the disappearance of manufacturing sites, the changes in the structure of the transportation and distribution system and the reorganisation of key urban utilities – notably gas, water and power generation – have led to considerable land-use changes. With the demise of the London docks in the late 1960s and 1970s, and the earlier total decline of the urban canal system in favour of rail and truck distribution, most of the warehouses in Docklands, and along the banks of the Thames and the Regent Canal, which rings central London from the Limehouse basin in the east to the City Road and King's Cross basins in the north and the Paddington basin in the west, lost their previous functions and fell into dereliction. Many of them are in the process of being converted into offices or apartments (Figure 9.2). The Beckton gasworks site (now called Gallions Reach) east of the Royal Albert Dock has been developed for housing, as has the old Rotherhithe docks (now called Surrey Quays), adjacent to Canada Water station on the Jubilee line to Canary Wharf. The former gasworks site at the tip of Greenwich peninsula was converted into the financially disastrous,

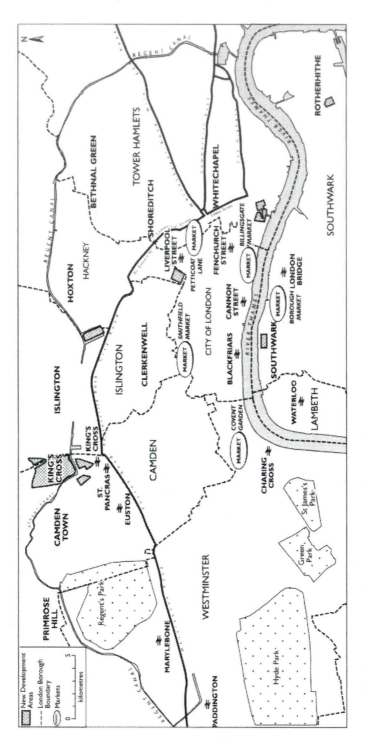

Figure 9.2 Major transport sites and markets in Inner London.
Source: Chris Hamnett.

though visually striking, Millennium Dome (Thornley, 2000; Pierce, 2002), and parts of the Deptford Creek area along the Thames are the focus of major urban regeneration projects. In November 2000 the giant Excel exhibition centre, with 90,000 square metres of space, opened overlooking the Royal Victoria Docks and another 65,000 square metres are due for completion by the end of 2003 (Waples, 2000). It will create an exhibition centre almost as large as the NEC in Birmingham. Docklands is the biggest single area of urban regeneration in western Europe and as such it is considered in detail below.

Nor should we forget the disappearance of the railway stations and goods yards surrounding central London. Broad Street station is now the Broadgate Centre, the vast semi-derelict area of King's Cross goods yards is likely to be redeveloped for offices and a new Eurostar station (Clement, 2000), and Paddington goods yard is soon to go. The former St Pancras station goods yard is now the new British Library, and all the major central London markets have now closed and been relocated or will soon be relocated: Covent Garden (vegetables and flowers), Spitalfields (fruit and vegetables), Smithfield (meat), Billingsgate (fish). Covent Garden is now one of London's most fashionable shopping and tourist areas, Spitalfields is scheduled for massive new development (Fainstein, 1994a; *The Economist*, 2001), and the old Billingsgate market by Tower Bridge was redeveloped for financial uses. Not only has land-use redundancy made these sites available, but in several cases their distinctive Victorian architecture has been a positive point for their redevelopment. Figure 9.2 provides a location map of many of these sites.

If we add to this list the miscellaneous collection of old industrial buildings and sites which have now been transformed into new functions, including the transformation of Bankside power station into Tate Modern, the conversion of the former Daily Express newspaper building on Fleet Street into a headquarters for Goldman Sachs, rebuilding the old Chelsea Harbour (yacht marina and luxury apartments), the conversion of the Harrods Furniture Depository in Barnes into luxury apartments and the Oxo Tower on the South Bank into a luxury restaurant and apartments, it is possible to see the gradual disappearance and transformation of large tracts of London's old industrial landscape into the building blocks of a new post-industrial landscape (Savitch, 1988; Williams, 1992; Zukin, 1992; Teedon, 2001). There are still a few relicts left (such as the remains of Battersea power station scheduled for conversion into an entertainment complex) owing to the financial vagaries of the property development industry and planners, but they too are likely to soon disappear. As Thornley notes of the Millennium Dome project:

> The Dome project may appear to be an exceptional and unusual example of urban development. However, this is not the case. It is illustrative of the transformations that are taking place in cities throughout the

world ... Such projects are the result of global economic and cultural changes ... In the post industrial city, economic survival depends upon growth in the service sector ... The Dome project fits into the trend towards leisure-orientated development. It seeks to attract large numbers of visitors to its spectacular entertainment experience, located on an old industrial stretch of the Thames.

(Thornley, 2000, p. 689)

The fact that the Dome was an economic failure almost from the start, and led to the government writing off approximately £650 million, should not detract from the fact that this general type of development is of growing importance as a way of trying to regenerate derelict or abandoned old industrial areas. In other areas of Docklands similar sites have been converted into a new campus for the University of East London, the large new exhibition centre (Excel), and, further downstream, giant retail centres such as Lakeside or Bluewater have been created in old quarries and the like. It would be a mistake to view these developments in isolation. They all form part of a consistent transformation from an industrial to a post-industrial land-use pattern. A similar process can be detected in part of the residential transformation of London.

The transformation of the residential landscapes of Inner London

When Steen Rasmussen (1934) wrote his classic study of the built form of London, *London: the Unique City*, he argued that it was physically distinctive from other European cities because of its low density, terraced and semi-detached housing. This is still partly true. But although large areas of Inner London still retain their Georgian, Victorian and Edwardian terraces, and Outer London remains predominantly semi-detached (Jackson, 1974), the post-war period has seen a number of changes which have reshaped the residential face of central and Inner London. This has taken two radically different forms. First, at the top end of the private housing market, there has been a transition from industrial to post-industrial landscapes with the conversion of large numbers of old warehouses, industrial and office buildings for luxury residential apartments. Second, large areas of older poor-quality housing have been redeveloped as high-density and high-rise council estates. We will discuss each of these in turn.

Private market conversion of industrial and commercial uses to luxury residential

This process has been marked in Docklands where large numbers of old buildings have been converted in addition to a large number of new luxury apartment buildings which have been built along the river. The conversion process started in Wapping just to the east of the City of London in the 1970s, but soon spread to the south bank of the Thames, with spectacular conversions of Butler's Wharf and other buildings. It was temporarily interrupted by the property market downturn of the first half of the 1990s, but the process subsequently spread to Clerkenwell and Shoreditch with a pioneering conversion by the Manhattan Loft Corporation (Hamnett and Whitelegg, 2001).

The conversion of industrial buildings is now widespread. In Clerkenwell, the former headquarters of the Metropolitan Water Board (New River Head) is now luxury flats, and Paton House and the former Dr Scholl footwear factory have been converted into flats, along with many other buildings. Warner Lofts have converted the Paramount Building and the Art Deco Ziggurat, and in Shoreditch, a former company headquarters, The Exchange Building is now luxury apartments; also, with a nice touch of post-industrial irony, the former Carreras cigarette factory on Mornington Crescent is now the head office of a major mass tourism package holiday company. The promotional material for some of these conversions is illuminating. The literature for the Ziggurat building states that:

> London's Art Deco landmark buildings have always been the great flagships of industrial architecture, uniquely popular as a commercial building style. They remain the undisputed 20th Century classics of commercial architecture . . . In the 90's, these types of buildings have been recognised as ideal structures for a new way of living, the New York loft-style apartment. Outside, a classic architectural form, inside a column based structure offering open, unencumbered spaces that allow people to realise their own vision of city living.

Nor is the conversion process confined to old warehouses and industrial buildings. It has also spread to a wide variety of older office buildings in both central London and the city fringe. A massive 1960s office block on the South Bank immediately behind the Hayward Gallery has been converted to 'the White House', and many other office blocks have been converted during the 1990s. Along High Holborn former insurance company buildings such as the Prudential building have been, or are in the process of being, converted into offices and apartments as the insurance companies have realised that they no longer need to do routine claims processing in expensive central locations and have decentralised the bulk of their operations. So extensive has the process been that Cadogan (1999) raised the question of 'What is there left to convert?' Many of the potentially convertible buildings in Clerkenwell have been converted, as have most of the riverside buildings in Docklands. The Harrods Furniture Depository on the south bank of the river at Barnes has been converted into 'Harrods Village' (the names are

Plate 9.1 Converted lofts in Shoreditch.
Source: Chris Hamnett.

always revealing), and almost every former riverside industrial site is now a luxury development. Estate agents suggest that developers are turning their attention to areas south of the river such as Borough and Bermondsey and are also looking further down river.

What London has seen is the 'revalorisation' of the riverside from the old industrial and port processing zone into a prestige residential area for luxury apartments. In this respect, it is moving back to the riverside palaces of the pre-industrial age when the Thames was lined with palaces of the king and court. The most spectacular proposed development is for a 49-storey residential tower complete with a 400-bedroom hotel, shops, restaurants and a fitness club on a derelict site in Vauxhall on the south bank of the Thames. The developers view it as part of the regeneration of the area, and the architects compare it to Trump Tower (Jameson, 2001c; Shah, 2001a). If it is built it will take its place near Albion Riverside, a spectacular luxury apartment development located between Albert Bridge and Battersea Bridge (Plate 9.2).

This process is, of course, common to all global cities. It is part of a widespread shift from industrial to post-industrial city-centre land uses. Although in the UK the process was primarily confined to London, in recent years it has expanded rapidly to Leeds, Manchester and other northern cities where there is a growing young service class and a market for conversion of old factories and warehouses into city-centre apartments.

The conversion of large numbers of old industrial and office properties did not happen out of thin air, however, and it is important to explore the complex reasons underlying conversion. It is suggested that five principal reasons help explain the sudden boom. First, the growth of the new middle class provided the demand base for development. As Chapter 7 argued, the

Plate 9.2 Albion Riverside.
Source: Hutchison Wampoa Property.

expanded new middle class has underpinned gentrification. Second, the transfer of the idea of loft living from New York had captured the public imagination, or certainly the imagination of the specific group of young, design-aware professionals (Zukin, 1988). Third, and crucially, the downturn in the office market in the early 1990s (which was particularly marked in the peripheral areas of the city) led to property owners and developers looking at alternative ways of profitably using their buildings. Fourth, the central government decision to change planning legislation in 1987 and designate the B2 development category of light industrial and office uses allowed developers to convert old factories into offices without planning approval. Last, but not least, the gradual change of policy by local planners and politicians to cease trying to defend manufacturing industry and permit change of use from business to residential allowed the process to get under way. Without these factors it would not have happened. The limit to the process, in London at least, appears to be the rapid rise in office rents at the end of the 1990s. Whereas in the early 1990s the developers were looking for profitable uses for old office buildings, rising office rents have raised the opportunity costs of conversions (Hamnett and Whitelegg, 2001).

Local authority housing: the other side of the coin

The second, and one of the most directly visible and striking transformations of the built environment of London, has been the replacement of large areas of poor-quality private rented housing by high-rise and high-density council estates. Driven by a very different set of factors, they were produced for a very different clientele. Although the former LCC had built some medium-rise blocks in the inter-war period, and a handful of pioneering developments in the 1890s, primarily in the East End, they were small scale and they focused their energy on building suburban housing estates where land costs were lower. Post-war council redevelopment was on a much larger scale.

As Chapter 6 pointed out, these large-scale council redevelopments were viewed in the 1960s and 1970s as a modernist panacea to poor housing conditions for the working-class population of Inner London. A combination of strong local Labour councils and Labour and Conservative central governments committed to large-scale subsidised social housing building programmes ensured that the transformation took place very rapidly (Dunleavy, 1981). Some of the estates, such as the ones built at Roehampton and some in Camden, were very well designed and well built, but many were quickly thrown up by large builders using prefabricated concrete panelled system building methods to cut costs and maximise profit (Merrett, 1988; Glendenning and Muthesias, 1994). Regrettably, many of them have steadily deteriorated from architectural and planning utopias in the 1960s to decaying, unsafe and unpopular sink estates with high vacancy rates (Harrison, 1983; Braid, 2000; Power, 1996). In some boroughs, such as Hackney, the worst blocks have been demolished to popular acclaim. In other cases, the estates have been remodelled by removal of high-level walkways, height reductions, the construction of low-rise housing and other strategies designed to make them more liveable (Coleman, 1985). The problem is not high-rise or high-density housing *per se*, but a combination of poor design, poor construction and widespread resident poverty. The Barbican development in the City of London is one of the most sought after and expensive residential areas in central London – but it is well built, well maintained and socially selective by income. In most areas, however, the physical legacy remains and most local authorities are so cash-starved that they are reliant on central government for funding for renovation or demolition and rebuilding of the worst estates. The high-rise solution to new housing was adopted only briefly in London, and GLC statistics show that even at its peak in the period 1965–71, 70 per cent of new council housing was provided in 1–4-storey buildings, compared to 30 per cent in buildings over 5 storeys, and only 6.4 per cent of total units were in buildings of over 10 storeys. Although the figures help put the phenomenon in perspective, Garside (1983) notes that they 'greatly understate the concentration of these high-rise flats in particular areas of Inner London' (p. 39). Along with the large numbers of high-density blocks, post-war council redevelopment has remodelled large areas of Inner London, often physically for the worst. Figure 9.3 shows the geographical extent of council estates in the inner areas of Lambeth.

Unlike many other aspects of the transformation of the built environment, the rise of the post-war council estates cannot be simply attributed to the power of capital. On the contrary, they are a direct result of central and local government housing and planning policy. However, as Dunleavy (1981) has demonstrated in his impressive study of the politics and finance of mass housing, they owe a great deal to government policy on subsidies to high-rise housing, to the desire of many construction companies to use cheap system building methods, and to resistance by the Outer London Conservative local authorities to council building within their boundaries (Young and Kramer, 1978; Saunders, 1979; Garside, 1983). In their determination to ensure that Outer London remained dominated by private housing and the middle classes they helped ensure that much of Inner London was redeveloped as high-density council housing.

The redevelopment and expansion of the City of London

The growth of financial and business services in London, and the strong demand for office space which has accompanied it, has been very important in the transformation of London (Coupland, 1992a). It has led to the gradual replacement of the old Victorian office buildings of the City of London with large new office buildings, the expansion of the City into peripheral areas,

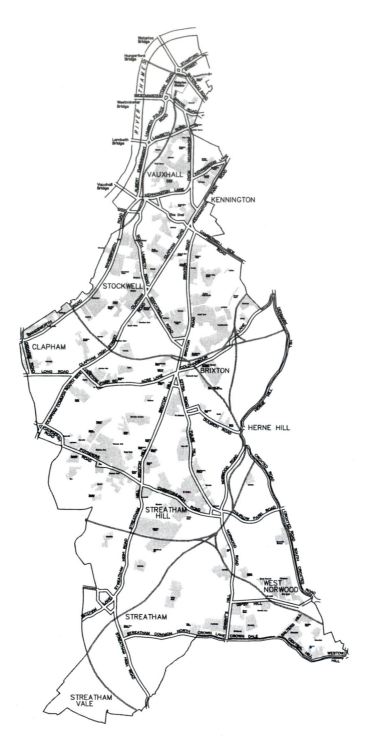

Figure 9.3 The geographical extent of council estates in inner Lambeth.
Source: Chris Hamnett.

and the creation of new office areas in Canary Wharf and elsewhere. Just as the previous expansion of the financial and commercial role of London in the Victorian era led to a wave of new building (Black, 1999, 2000), so the period of post-war reconstruction and the impact of Big Bang and its aftermath in the late 1980s and the 1990s have led to a wave of new building.

Traditionally the banking, insurance and trading activities of London were contiguous with the administrative boundaries of the historic 'square mile' of the City of London (indeed they occupied only a small fragment of the City of London – see Figure 9.4). One reason for this was the floor-based nature of trading and the necessity for traders and their back offices to be housed within easy walking distance of the trading floors. It has only been with development of improved telecommunications and linked dealing screens and electronic markets that it has been possible for floor-based trading to be replaced by electronic trading. This started on the Stock Exchange in 1995 and it only replaced floor trading in LIFFE in 1997. In recent years, many streets in the City and central London have been torn up to allow deployment of networks of high bandwidth fibre optic cables which allow very high volumes of electronic data to be transmitted and received instantaneously (Graham, 1999; Moss, 1987). The development of new electronic technologies has not only permitted a greater degree of decentralisation, but, in conjunction with the changes in corporate organisation, it has forced the pace of change in building design and requirements.

The rise of the new, mega firms in the City such as Deutsche Bank, CSFB, Goldman Sachs, Merrill Lynch, and HSBC, with large foreign exchange, derivatives and stock market dealing teams, has led to the need for new buildings with large new, open-plan, dealing floors. Because of the almost total shift to an electronic trading environment, with each dealer being surrounded by a battery of screens showing trades and prices, this has had major implications, not just for the overall amount of space required but for the type of space itself. In order to accommodate these open-plan dealing floors, large new buildings are required, and to accommodate all the cabling, ducting and air conditioning required to service the electronics, these dealing floors need a minimum 10 feet of height between floors. This is rarely found within existing buildings and changing technological requirements have led to a need for new buildings. In addition, mergers within the banking, legal and consulting sectors have generated pressure to centralise operations in new buildings. Such buildings either did not exist or were already let. As a result, there has been a demand for new corporate headquarters. This has driven the demand for large-scale office construction in the 1980s and 1990s both within the City and in its immediate surroundings and more widely (Figure 9.4).

The expansion of the City out of its traditional core into its adjacent fringe areas was relatively slow at first but, as Pryke (1991, 1994) has pointed out, one consequence of Big Bang and the employment growth of the City in the

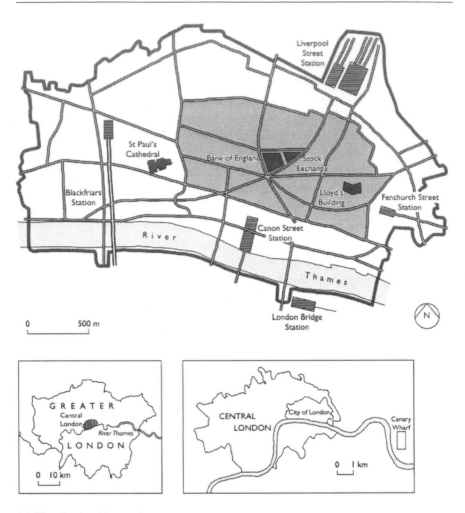

(a) The City's old spatial matrix.
Figure 9.4 The expansion of the City of London.

1980s and 1990s, combined with the operation of tight planning controls upon construction of high buildings, has been the physical expansion of the main financial district beyond the historic confines of the City of London. One of the first manifestations of this was the development of the Broadgate centre next to Liverpool Street station at the northern tip of the City in 1985. At the time, the Broadgate Centre represented the largest planning permission in the history of London, with fourteen buildings on 29 acres of formerly disused railway land. But Williams (1992) identifies a number of other similar developments, which she terms 'groundscapers', rather than skyscrapers, because of their relatively low-rise but spatially extensive nature.

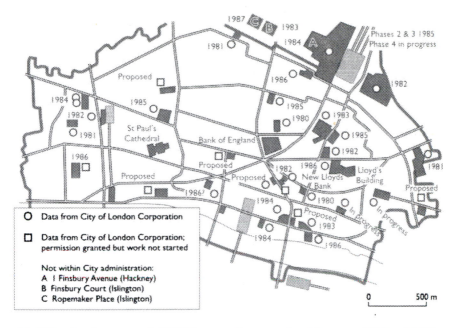

(b) Office developments of 100,000 square feet completed between 1979 and 1986 or proposed as at December 1985.

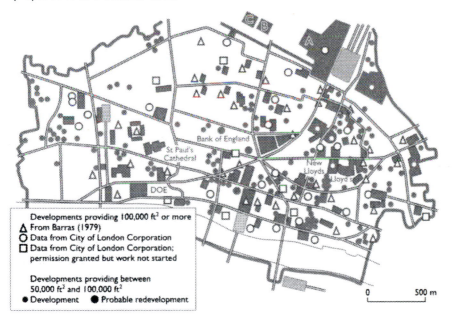

(c) Office developments completed between 1960 and 1986.
Source: M. Pryke (1994).
Figure 9.4 Continued

She argues that the kind of requirements now seen as necessary for the modern financial institution, including dealing rooms, flexible office space and the like which emerged in the run up to and subsequent to Big Bang, 'do not allow for old fashioned infilling. They need a whole city block' (p. 248). But given the opposition to high-rise buildings, the logic of development has given rise to dense, deep-plan buildings that cover every inch of their site. Williams suggests that what is taking place 'represents the biggest remaking of London since the Second World War' (p. 248).

The changing development stance of the City of London

The reshaping of the City of London and its immediate environs has been a relatively slow process, and many of the early post-war office redevelopments such as London Wall, the Elephant and Castle and Paternoster Square can be seen, in retrospect, to be rather crude, uninspired and unattractive exercises in modernist architecture designed to yield maximum space as cheaply and as quickly as possible. They were the office equivalents of high-rise and high-density mass council housing. Fortunately, many of the more recent developments such as the Broadgate Centre, the Charing Cross station redevelopment (Embankment Place), No. 1 Temple Place on the Embankment and the new Paternoster Square have been built with greater commitment to good design.

The redevelopment of the City has not been without considerable conflict. The City of London Corporation planning department struggled until relatively recently to contain new developments and to maintain the traditional landscape of the City in the face of pressure for high-density and high-rise redevelopment. Peter Palumbo's long struggle to gain planning permission for a prestige site in the heart of the City, No. 1 Poultry, opposite the Bank of England, has been comprehensively documented by Jacobs (1992, 1994). Another major battle has been over the proposed redevelopment of Spitalfields market on the north-eastern fringe of the City. The Corporation of London is the freeholder of the land and was keen to demolish and redevelop the western half of the site (the eastern half is on the preservation list) as a ten-storey 120,000 sq. ft office development. It was opposed by local preservationist groups and the Spitalfields Small Business Association who wanted Tower Hamlets council to refuse planning permission on the grounds that the market is a lively contribution to the local economy (see Forman (1989), Fainstein (1994a) and *The Economist* (2001) for discussion of the plan). After 13 years of struggle, the council gave permission in October 2002 and construction is under way.

On the other side of the fence the City of London Corporation and business pressure groups argue that it is vital for the development to go ahead to ensure that the City retains its world-class status and does not lose out in the struggle to Frankfurt. This, of course, is questionable, as it may be the existence of things like Spitalfields market which are part of the attraction of London as a place to live and work (Jenkins, 2000).

What is interesting about recent debates, however, is the fact that the City now seems to have come down firmly in favour of much more intensive development within its boundaries in the face of competition from both Frankfurt and from Canary Wharf. In this respect, traditional planning criteria appear to have been changed to match the changing requirements of continuing financial dominance. Mathiason (2001), in an article headlined 'City set for huge expansion', noted that:

> The Corporation of London is planning a dramatic expansion of the Square Mile to cope with huge demand from international banks and to maintain the City's place as Europe's top financial centre. The move would see 10 million square feet of offices added – about 12.5% of the current total.

The report stated that the City Corporation (the City's local authority and principal landowner) has earmarked three sites for expansion and has bought land so it can force through its expansion plans. One is a 10-acre site next to the Spitalfields market in the neighbouring borough of Tower Hamlets which will accommodate towers and buildings totalling 4 million square feet. The second is on the west side of Spitalfields, and the third market is on a site close to the Barbican and close to Farringdon station in Clerkenwell which is adjacent to good transport links.

In many ways, this marks a considerable turnaround, particularly for the Corporation of the City of London who, after many decades of balancing the protection of the traditional City landscape against the demand for more office space, suddenly saw the departure of several major financial firms to Canary Wharf in the late 1990s. This concentrated the City's mind on the need to shift the balance in favour of more offices if they were not to lose their traditional dominance. It is significant that the struggle over the redevelopment of the Baltic Exchange, which was damaged by an IRA bomb in 1994, has been resolved with the granting of planning permission for a major 41-storey high-rise for insurance giant Swiss Re (see Plate 9.3) (Shah, 2000), which may be indicative of a change of planning stance in favour of high-rise development. July 2002 also saw approval granted for the 37-storey Heron Tower at 110 Bishopsgate (Davey, 2002c). Other applications have been submitted for a 34-storey tower in Fenchurch Street, and London Bridge, immediately south of the City, is also the proposed site of a 420-metre-high tower which would dwarf the 244-metre-high Canary Wharf tower (Waples, 2002).

Some would argue that it is too late, and that the development horse has bolted down river to Canary Wharf, but the City will probably be able to reclaim the situation if it gets government support for a policy of selective

Plate 9.3 The 'erotic gherkin': the Swiss Re Building, under construction.
Source: Chris Hamnett.

high-rise development. In retrospect, Canary Wharf may be seen as necessary to ensure London's continuing domination of the European financial market in that it has provided a large amount of high-quality, modern, office space which was not, and is not, available in the City (Davey, 2002b). For developers the key problem is getting planning applications decided quickly.

The high-rise offices debate

London had relatively few high-rise office towers until the development of Canary Wharf. The highest one in the City is Tower 42 (previously the Nat West tower). The City of London Corporation was generally opposed to high buildings. However, the growing strength of Frankfurt, high and rising office rents in London, and the demand for large buildings by financial institutions has led to the reassessment of this policy.

Even Ken Livingstone, the mayor of London, has now become a strong proponent of clusters of high-rise office towers around transport nodes if London is to compete effectively with other major European financial centres and maintain its dominant role as one of the world's leading global cities:

> This is an argument about whether London stays as a world city capable of delivering the employment, investment, infrastructure, housing and office space that are needed to compete with New York, Tokyo, Frankfurt or Berlin . . . London isn't Manhattan. What I envisage is a small number of tall buildings on locations suitable for their development, to meet an identified economic need and to contribute to London's overall vitality as a world city.
>
> (Livingstone, 2001b)

Needless to say, these proposals have generated strong opposition from those who think that more high-rise office towers are unnecessary and will destroy the London skyline (Blackhurst, 2001; Binney, 2001). As Simon Jenkins put it:

> Central London has not the remotest need of towers. Its texture is not that of hard-edged Manhattan . . . Livingstone's vision of the London skyline is incoherent, a desert strewn at random with Ozymandian stumps of buildings.
>
> (Jenkins, 2001)

Ken Livingstone disagrees, pointing out that:

> I broadly expect clusters of new higher buildings in the north eastern corner of the City boundaries, on some of the major London rail terminuses, such as London Bridge, and in other areas such as Canary Wharf

> ... I will listen to those who oppose new tall buildings but at least recognise the requirement to compete for business and investment with New York, Tokyo, Berlin or Frankfurt. Those who simply refuse to recognise there is an issue at all fail to connect with the issues London is confronting. The Corporation of the City of London is aware of 14 companies wishing to locate in London in buildings of at least 500,000 square feet and further demand is evident at Canary Wharf.
>
> (Livingstone, 2001b)

The issues raised regarding high towers are, in many ways, simply a new version of the long-standing debates regarding the need for more modern office space in central London if it is to compete with other cities. The issue is about the extent to which it is necessary to adjust London's built form to adapt to the needs of contemporary global capitalism. Livingstone has decided that it is, and one of his first pronouncements on becoming mayor (Livingstone, 2001a) was on the need to ensure that London remains competitive with other major global cities. He has undergone a transformation from a populist opponent of office developers when he was leader of the GLC in the 1980s to recognising the need for development to maintain London's position and its jobs (Sudjic, 2001). He has, however, also strongly reiterated the importance of providing more affordable housing in London for key workers if London's services are to be able to function properly and is pressing for at least 25 per cent in all new developments.

The redevelopment of Docklands

The one development which has arguably changed the face of London more than any other is the redevelopment of Docklands and, in particular, the emergence of Canary Wharf as a major new financial and office centre. The history of Docklands – which runs from St Katherine's Dock immediately east of the Tower of London in the west to the Royal Docks in the east – is well known and reflects London's rise in imperial and commercial importance. The first to be built was London Dock in 1805, followed by the East India Dock in 1806 and St Katherine's Dock in 1828. This was followed by a further round of dock building in mid century, to cater for steamships and larger vessels. The Victoria Dock was built in 1855, the Surreys in 1858, Millwall in 1868, the Royal Albert in 1880, and Tilbury in 1886 (Brownhill, 1990). Subsequent over-capacity of docks and warehouses led to the Port of London Authority (PLA) being set up by Parliament in 1909 to take public control of the management of the docks, and the last docks in London to be opened were the King George Fifth in 1921 and the Quebec in 1926 (Figure 9.5) (Brownhill, 1990).

The decline of Docklands began in the late 1960s, driven by containerisation, the increasing size of ships, a general decline in world shipping and

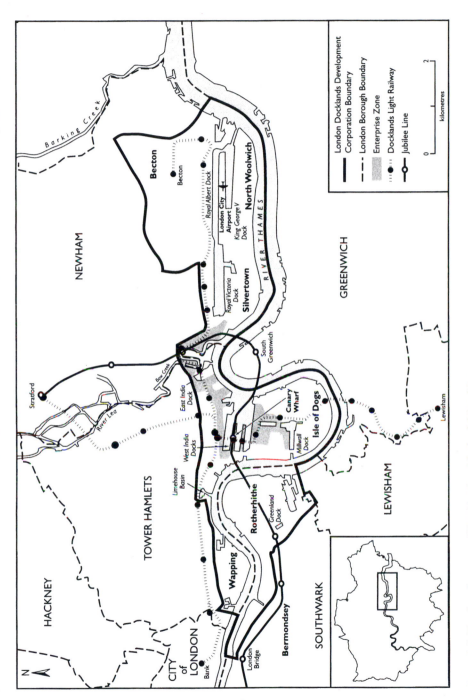

Figure 9.5 The key features of Docklands with the Jubilee Line.

Legend:

- London Docklands Development Corporation Boundary
- London Borough Boundary
- Enterprise Zone
- Docklands Light Railway
- Jubilee Line

0   1   2
kilometres

HACKNEY

NEWHAM

Becton
Becton

North Woolwich

London City Airport
King George V Dock
Royal Albert Dock

Royal Victoria Dock

Silvertown

RIVER THAMES

GREENWICH

CITY of LONDON

Bank

Stratford

River Lea

Barking Creek

Bow Creek

East India Dock

West India Docks

Limehouse Basin

TOWER HAMLETS

Wapping

London Bridge

Bermondsey

Rotherhithe

Greenland Dock

SOUTHWARK

Canary Wharf

Isle of Dogs

Millwall Dock

South Greenwich

LEWISHAM

Lewisham

N

construction of new container ports such as Felixstowe. The East India Docks were closed in 1967, the London and St Katherine's in 1968, The Surreys in 1970, the West India and Millwall in 1980 and the Royals in 1982, with the PLA concentrating its shipping activities in Tilbury, the newest and furthest downstream of all the docks. The decline of the docks was accompanied by the decline of port processing and manufacturing industries, and between 1961 and 1971 83,000 jobs were lost in the five boroughs of the Docklands area: Greenwich, Lewisham, Newham, Tower Hamlets and Southwark. In 1967, just before the first closure, there were still 23,000 registered dock-workers in the Port of London. By 1980 this had fallen to 4,000 and in 1981 the closure of the Royal docks was announced. The subsequent growth of unemployment was accompanied by population decline. Tower Hamlets lost 18 per cent of its population between 1961 and 1971 compared to an Inner London average of 10 per cent (Church, 1988). The dock era lasted about 160 years in all, but by 1981 it had ended and large tracts of East London lay derelict. The question then emerged as to what should be done with Docklands.

Redeveloping Docklands: the early stages

The first dock to be redeveloped was St Katherine's dock, nearest the City of London. The PLA sold the site to the GLC for £1.25 million in 1968, and in 1969 the Conservative GLC initiated a competition among private developers based on a development brief for a mix of commercial and com-munity uses. The winning proposal included offices, a World Trade Centre, a hotel, a mixture of council and private housing and shops, and a private yacht marina. The development proved a commercial success, but Brownhill argues that 'although it was supposed to be a mixture of high-value and com-munity uses, the emphasis on private-sector-led proposals, commercial devel-opments and minimal public involvement was to set the scene for much of what was to follow' (1990, p. 21).

The next, and crucial, step was initiated by Peter Walker, the Conservative Secretary of State for the Environment, who announced in 1971 that a com-prehensive study of Docklands was to be carried out by the private consult-ants, Travers Morgan. Walker stated that his explicit aim was to replace Docklands' declining industries and to help 'bring the West End into the East End'. Travers Morgan came up with a wide range of proposals ranging from one based on industry and public housing to the favoured one – 'City New Town', which would involve 60,000 new office jobs, 1,000 industrial jobs, and a population of 100,000 housed 50/50 in private and public housing. There was a definite emphasis on changing the social mix of the area to ensure more middle-class residents, and redevelopment was to be based on large-scale commercial and office development with an emphasis on private sector investment (Brownhill, 1990).

The proposals, which involved 5,500 acres of land, and which were overseen by a steering group involving central government, the GLC and the five boroughs whose boundaries overlapped the area, but no local politicians, generated considerable local opposition and led to the formation of the Joint Docklands Action Group (JDAG) and the North Southwark Community Action Group: 'the basis to the opposition was that the Travers Morgan proposals took as the starting point not the interests and needs of the local population or the boroughs, but rather, those of central government and the private sector' (Brownhill, 1990, p. 23). The boroughs were hostile because they could not use their planning powers and because of their fears that the government wanted to set up a New Town type of authority to implement the proposals. In 1974 central government accepted the borough and GLC arguments for a new planning body and established the Docklands Joint Committee with the task of producing a strategic plan for the area.

The election of a Labour government in 1974 and the establishment of a Docklands Forum, which involved forty locally based groups, also helped switch control of Docklands planning back from central government to the local state. Work started on production of the London Docklands Strategic Plan in 1974 and a draft plan was produced for public consultation in 1976. The plan was radically different from the Travers Morgan plan. As Brownhill comments:

> Economic policies were not based, as in Travers Morgan, on the assumption that new office and service sector jobs would reverse the area's decline. Instead, industrial employment had to be stabilised and new industry attracted, with service employment being a secondary contributor. The docks which were still open in 1976 were assumed to remain operational, and to form a major part of the economy . . . Housing proposals centred on plans to build 23,000 new houses, 50% for rent, 30% shared ownership and 20% owner occupied with emphasis on rented accommodation for the early years.
>
> (Brownhill, 1990, p. 26)

Brownhill notes, however, that 'Producing a plan was one thing, implementing it was quite another' (p. 26). The problems the DJC faced were twofold: land availability and second finance. The plan was estimated to cost £1,138 million at 1976 prices. But government had already stated that no additional money was likely to be forthcoming; then, in 1976, following a financial crisis, the IMF imposed dramatic expenditure cuts on the Labour government. The London docks and the Surrey docks were acquired and filled by the GLC and the boroughs, but much of the land earmarked for development in the plan was owned by public bodies such as the PLA and the Gas Board. The DJC hoped that they would be able to assemble the necessary land banks, but little progress was made – not least because existing landowners valued their land on the basis of either its hope value or historic cost. By 1979 both the DJC and the LDSP were in difficulty. Despite some progress there were significant shortfalls in both industrial jobs and housing production. Brownhill notes that 'by 1981 only 800 jobs had been created with 8,500 lost. This compares with a target of 10–12,000 new jobs by 1982' (1990, p. 30). It was against this background that Michael Heseltine, the Secretary of State for Environment in the newly elected Conservative government, announced his intention to set up Urban Development Corporations (UDCs) in Merseyside and London Docklands.

The UDCs were given significant land acquisition and planning powers, including the power to have the land within their boundaries 'vested' in them without recourse to any appeal, and to give planning permission for development with consultation or planning inquiries. In addition, they were accountable only to Parliament and were run by a board appointed by the Secretary of State and not by local representatives. Essentially, they were designed to take major development out of the hands of local politicians, local authorities and local interests, which were deemed too parochial and inward looking to ensure effective development. Government argued that because of its location close to the City, Docklands should be developed in the national and not the local interest and that what was needed in Docklands was 'the funds and energies of the private sector'. Government took the view that only independent development corporations could be flexible towards the private sector, give speedy decisions and change the market's perception of the area from one of risk to opportunity. The bill for the UDCs allowed a House of Lord's select committee inquiry to hear arguments against the government's proposals, but the committee approved the government's bill arguing that nothing would be done to halt the decline of Docklands unless there was 'a change of approach – a change of priorities'. Private investment and housing were needed and the LDDC was more likely to attract this than the DJC. Thus the LDDC was established in July 1981 and assumed its planning powers two months later. The emphasis was now firmly on the key role of the private sector in urban regeneration (Church, 1988; A. Smith, 1989; Foster, 1999).

The rise of the London Docklands Development Corporation

The task given to the LDDC was to 'regenerate' Docklands and it set itself the following objectives:

- To rapidly improve the image of Docklands through physical works and creating confidence in the continuing improvements to come;
- Since the amount of public money is limited, to use its resources primarily as a lever to attract private investment;

- Where prime sites are not the subject of suitable active redevelopment, to acquire as much of such land as resources permit in order to undertake the necessary reclamation, servicing and site assembly, followed by re-marketing to the private sector;
- To undertake major environmental works;
- To publicise and sell the opportunities of Docklands, as well as to provide potential customers with the information they require;
- To bring about significant improvements in the choice and quality of housing and in community amenities, but without undertaking such work directly;
- To bring the roads and public transport network up to the standards enjoyed in other parts of London;
- Finally, to wind up the corporation as soon as Docklands prosperity has clearly become a self-generating activity.

(LDDC, 1982, p. 6)

It is clear from these objectives that the LDDC saw its primary role as being to use public funding as a lever to attract private funding. Its strategy for doing this was by land acquisition and servicing, transport improvements, and the creation of a positive image for Docklands by intensive marketing. This should not be surprising, given that the LDDC was initially chaired by Sir Nigel Broakes, a private property developer.

The importance of landownership as a key element in this property-led regeneration strategy was crucial. The 1982/3 LDDC Annual Report stated that 'The ownership of land and its orderly preparation and disposal therefore remain the basic keys with which to achieve the changes necessary for the permanent, speedy regeneration of Docklands.' The LDDC rapidly acquired a large land bank – some through land vested in it, some via compulsory purchase.

Major sites acquired included the 132-acre Southwark site in March 1983, the Royal Victoria Dock (360 acres) in 1984 and the 745-acre Royal Docks in 1985. By March 1988 the LDDC had acquired 1,975 acres of which 1,550 acres were land and 425 acres were water, and held a land bank of 1,316 acres. This amounted to 34 per cent of the total land area and 40 per cent of the combined land and water area (Brownhill, 1990). Much of the land acquisition was from public bodies, and the LDDC undertook extensive reclamation and servicing work including clearing polluted sites, strengthening river walls, installing services such as drainage, water, sewage, electricity and fibre optics.

By 1989 the LDDC had spent a total of £790 million of public money, of which 84 per cent had been spent on 'pump-priming', through expenditure on transport, land, marketing and other infrastructure works. Spending on local industry, housing and the community amounted to just 5 per cent. Brownhill argues that:

although the rhetoric behind the LDDC was that private sector activity and the rolling back of the state were the way to regenerate the inner city, the reality was different . . . Far from planning disappearing, substantial intervention in the land and property markets remained alongside large scale public investment.

(Brownhill, 1990, p. 380)

The rise, fall and rebirth of Canary Wharf

The development of Canary Wharf to its current status as the biggest single office centre outside the City is a remarkable story of triumph over adversity. Canary Wharf is located at the northern end of the Isle of Dogs, and the original scheme put forward in summer 1985 was for a new office centre comprising 8.8 million square feet of offices, hotels, shops, restaurants and 8,000 car parking spaces. The most controversial element was for three 60-storey office towers. The promoters consisted of a consortium of American banks, including Credit Suisse-First Boston and Morgan Stanley, and the development corporation Tramell Crow. The rationale behind the scheme was the growing demand for new office space generated by Big Bang, and the threat posed to the City by Paris and Frankfurt. When the initial backers pulled out, the scheme was taken over by Paul Reichmann's Toronto-based development company Olympia and York which had built the World Financial Center in New York. But Cohen points out that the

> tradition minded City was anything but enthusiastic. Early in the 1980s, top-hatted clerks from the discount houses walked the length of Lombard Street each morning, calling at every bank along the way. The London stock exchange still had a floor and all its members were on it. To suggest that business could be transacted remotely and anonymously three miles from Threadneedle Street, the City's centre and home of the Bank of England, was to overturn 300 years of history.

(Cohen, 1999)

Reichmann pushed forward rapidly with the development, making extensive use of its buildings as security for bank loans to finance construction. By the time the first tower, 'One Canada Square', was completed in early 1990 the office market in London had turned down sharply in the recession and tenants were difficult to find for what was then seen as a peripheral location with poor transport facilities – the Docklands Light Railway was the only form of public transport – and in order to attract tenants Olympia and York were forced to let space at extremely low rents. In addition, the value of the buildings on which the construction loans were secured had fallen sharply and Olympia and York had also diversified into other areas of activity in North America which had performed badly. The result was that the lenders took fright and called in their loans and Olympia and York collapsed into receivership.

Fortunately the receiver kept the development going and continued to market space to tenants. Canary Wharf came out of administration in 1993 and, in a remarkable twist, was sold by the banks in December 1995 to a consortium of investors headed by none other than Paul Reichmann for £800 million. By this time the Jubilee line extension connecting Waterloo and London Bridge stations to Docklands was under construction and the property market was turning up strongly with a shortage of quality space and rapidly rising rents. Canary Wharf began to attract large numbers of blue chip tenants. Canary Wharf began construction on two more towers: one as the international headquarters for HSBC and the other for Citibank. In April 1999 Canary Wharf floated on the London stock market, raising £500 million by selling 25 per cent of its equity. Since then it has gone from strength to strength, attracting growing numbers of major tenants. It is generally now regarded as an extension of the City (Daniels and Bobe, 1993), although there has been some bitter competition along the way with the City initially seeing it as a threat to the City of London.

Canary Wharf has three big investment banks as tenants, CSFB, Morgan Stanley and Schroder, Salomon Smith Barney, and is soon to accommodate HSBC and Citibank, both of which have their own towers under construction. In addition to the two extra towers, three further buildings are under offer to Clifford Chance (one of the largest legal firms in London), the Northern Trust Company and Morgan Stanley who already have 1 million square feet in three buildings. Work on foundations for all five buildings was launched simultaneously to fast-track construction when lettings were secured (Mortishead, 2000a; Jameson, 2000a). In October 2000 Canary Wharf Group entered into the FSTE 100 companies as the largest property company in Britain. In September 2000 Canary Wharf Group announced its first full-year profits (£54 million after tax), and Paul Reichmann stated that 'This year has been the most successful year in our history. The level of leasing activity has been the highest achieved in any single year and rents have reached new highs.' CWG has 6.2 million square feet under construction, which will more than double the existing 4.8 million square feet. The 11 million square feet development will eventually provide space for 90,000 workers (Shah, 2001d). Tables 9.1 and 9.2 show layout of the site with the major buildings and tenants.

Relations with the City, once a bitter rival to Docklands, are now said to be very good with Canary Wharf providing a vital spillover for the cramped Square Mile. George Iacobescu, the chief executive, claimed that 'If it had not been for Canary Wharf, the City would have lost a lot of financial institutions.' This may be true, but it is difficult to know whether financial institutions would have left London given its importance.

Table 9.1 Canary Wharf: major buildings, area and principal tenants

Property address	App NIA Sq ft	% leased	Principal tenants
1 Westferry Circus	219,000	100	Chevron Texaco, CSFB
7 Westferry Circus	179,300	100	EDS, EMEA, Edward S Jones
15 Westferry Circus	171,300	100	Morgan Stanley
17 Columbus Courtyard	199,500	100	CSFB
10 Cabot Square	636,600	100	Barclays Capital, WPP Group
20 Cabot Square	558,400	100	Morgan Stanley, Barclays Capital
One Canada Square	1,246,600	98	Daily Telegraph, KPMG, Mirror Group Newspapers, State Street Bank, Bear Stearns, Bank of New York
25 Canada Square	1,223,500	100	Citigroup
33 Canada Square	562,700	100	Citigroup
25 North Colonnade	363,200	100	Financial Services Authority
30 South Colonnade	296,100	100	London Underground
50 Bank Street	213,800	72	The Northern Trust Company
Total	6,044,100	98.7	
8 Canada Square	1,100,000		Sold to HSBC in April 2002

Table 9.2 Properties under construction at 30 June 2002 and agreed tenants

Property address	App NIA	Completion date	Major tenants
5 Canada Square	516,600	September 2002	
1 Churchill Place	1,000,000	July 2004	(Barclays Bank)
20 Canada Square	529,000	December 2002	
20 Bank Street	535,000	May 2003	
25–30 Bank Street	1,008,500	August 2003	
40 Bank Street	607,400	March 2003	(Skadden, Arps, Slate, Meager & Flom; Allen and Overy)
10 Upper Bank Street	1,000,000	July 2003	(Clifford Chance)
Total	5,527,000		

Source: Canary Wharf Group Preliminary Results for year to June 2002: Regulatory News Service 0815B, 12 September 2002.

Note: NIA = net internal area.

It is argued however that Canary Wharf has become a victim of its own success, with rents creeping up to levels comparable to those in Victoria in the West End. They are also facing competition from other large developments including Paternoster Square, adjacent to St Paul's Cathedral, and the second phase of London Bridge and Paddington Basin. Pearman (2000) suggested that 'Paddington is five million square feet close to the Heathrow Express link and it will be priced at competitive prices. The West End of London is where investment bankers want to live: Kensington and Notting Hill.' But as of April 2003 the downturn in the central London office market has led to the second phase of the Paddington redevelopment being put on hold. Also, the revelation that CWG have break clauses in their leases with tenants which permit them to return unwanted space has led to a sharp fall in the share price.

Canary Wharf: an appraisal

There has been a sharp difference in assessments of the success of Canary Wharf. The early 1990s academic and media consensus was that Canary Wharf was a financial disaster and a failed exercise in social engineering (Goodwin, 1991; Hebbert, 1992; Brownhill, 1990; A. Smith, 1989; Foster, 1999). Merrifield (1993) suggested that the financial collapse of Olympia and York and what seemed in the early 1990s to be the failure of Canary Wharf: 'could, perhaps, signal the end of the speculative, get-rich-quick, Thatcherite–Reaganite style of capitalism the 1980s came to epitomise' (p. 1247). A more recent appraisal (Florio and Brownhill, 2000) argued that although the LDDC has successfully transformed the physical landscape of Docklands from a derelict area to a desirable office address, it has increased polarisation rather than promoting social regeneration. They also argue that it was fundamentally anti-democratic and that the property-led planning 'strategy' did no more than try to hook together the initiatives being taken by a variety of private companies, quangos and state agencies. They do, however, concede that the over-supply of office property which was predicted by the critics of Canary Wharf in the early 1990s when vacancy rates were 60 per cent has proved to be a temporary phenomenon. Today, the vacancy rate is under 5 per cent, though it is estimated it could rise to 11 per cent or more if key tenants exercise their break clauses.

The radical academic view is totally at variance with recent media coverage which sees Canary Wharf as a great success and a triumph over near disaster. As Wheatcroft noted in *The Times* on the announcement of Canary Wharf's first profits:

> Canary Wharf is a stunning example – architecturally and financially – of how success can be built from disaster. Eight years ago, the receivers were marching in, yesterday, the company reported its first annual set of profits. The once-derelict Docklands site has become a mecca for investment banks and those who feed off them. What once seemed in danger of being a white elephant on the river is now such a popular location that it will soon be running out of space.
>
> (Wheatcroft, 2000)

Similarly, Norma Cohen (2000), the property correspondent of the *Financial Times*, in an article entitled 'The remarkable fall and rise of Canary Wharf', stated that:

> Of all the boom–bust–boom stories of the past few decades, Canary Wharf is one of the most spectacular. Last week, the company that owns the 85 acre estate in what was once the most desolate part of East London's Docklands, made it into the FTSE 100 index of Britain's biggest quoted companies. Yet just five years ago, Canary Wharf was owned by a group of bank creditors, having emerged from administration in 1993 with its obituary written many times over. Even as recently as March 1999, at the time of its flotation on the London Stock Exchange, its detractors were loud, many and influential . . .
>
> (Cohen, 2000, p. 20)

Who is right? In some respects, the question is unfair in that it juxtaposes academic articles written in the early 1990s, when the scheme was clearly in deep trouble, with recent press articles written when the success of the scheme seemed assured. But in another way there is much deeper conflict. There is no doubt that in terms of physical redevelopment of a once-derelict area, Canary Wharf and Docklands redevelopment more generally has been a dramatic success. Empty warehouses, old wharves and a decaying environment have been replaced with a gleaming new financial centre and large numbers of new-built or converted luxury dwellings. Until 2002 major financial institutions were queuing to take space, the developers were busy building more towers for prestige tenants who wished to make Canary Wharf their London, European or international headquarters and rents were rising very quickly. There is also no doubt that Canary Wharf has led to the relocation (though not necessarily creation) of tens of thousands of jobs in finance and business services to the area. The Docklands redevelopment has also led to the building of thousands of new dwellings in what was a hitherto derelict area. As Rhodes and Tyler comment in a summary of their consultancy report evaluating the regeneration of Docklands:

> The LDDC has successfully tackled the widespread multiple market failure that prevailed in London Docklands in 1981. Failures in land, housing and commercial property markets have been addressed and labour market failures have been alleviated by a combination of training projects, improvement in accessibility in and out of Docklands and the creation of new local jobs . . . With respect to value-for-money, the evaluation concluded that every £1 million of public sector cost generated net additional benefits in the UDA of 23 jobs, 8300 sq m. of office floorspace, 7.8 housing units plus many other diverse and intermediate benefits.
>
> (Rhodes and Tyler, 1998, p. 32)

The question which vexes many radical academics and planners, however, is to what extent local residents have benefited. For them Canary Wharf in particular, and the Docklands redevelopment as a whole, represents the triumph of global capitalism over local democracy and the needs of local residents, who have been systematically ignored. The radical argument, expressed most clearly by Goodwin (1991), Brownhill (1990) and Foster (1999), is that Docklands involved the removal of local democratic control and the replacement of one set of jobs and residents by another very different set. Goodwin (1991) suggests that Docklands can be seen in terms of the 'replacement' of a surplus population. There is little doubt that he is correct in this. The objective of the LDDC was *not* to provide more council housing for local people. On the contrary, it would probably have been argued within the LDDC that the problem of the area was that with 85 per cent of households in Tower Hamlets in the council sector, that there was too much council housing rather than not enough. Their underlying social objective was to develop the private sector of the housing market in order to attract in a new and more prosperous set of residents to match the new jobs being created. In this respect the LDDC have been remarkably successful. A total of 24,000 new housing units have been built, almost all of them for owner-occupation or private renting, and the housing tenure mix of the area is scheduled to move from 85 per cent council housing and 15 per cent private to a 50:50 ratio. While a small amount of social housing has been built, the LDDC's overriding objective was to attract private developers and house builders to increase the size of the owner-occupied sector. In this they have been very successful. And, given that Tower Hamlets witnessed the greatest percentage increase in the size of the professional and managerial socio-economic groups 1981–91, they can be said to have been very successful in replacing a 'surplus population' with new groups of young urban profes-sionals in the 1980s and 1990s (Short, 1989). It is illustrative of the nature of the new developments that an advertisement in the *Financial Times* in November 2001, which was marketing apartments in New Providence Wharf, quoted Skidmore, Owings and Merrill, the architectural and planning con-sultancy, as saying: 'After master planning Canary Wharf, our next challenge was to create a new urban quarter for Docklands.' The development includes 735 riverside apartments set in extensive grounds and with one-bed apart-ments priced from £294,950 and two-bed from £359,950. The development includes 'a 400 bedroom 5 star hotel, 500,000 sq ft of offices, retail, restaurants and state of the art leisure facilities'.

Both the radical charges are essentially correct. The very establishment of the LDDC was designed to go over the heads of the Labour-controlled local authorities and put redevelopment in the hands of a commercially orientated body with an international vision which would not be hamstrung by local planners or politicians pushing for the needs of local residents and local jobs. Indeed, the LDDC would probably argue that this was precisely

why the redevelopment had to be taken out of the control of local, parochial interests and put in the hands of an organisation with a far bigger
vision.

Had the local councils succeeded in taking control of the redevelopment it
is likely that the form of the redevelopment would be very different, much
smaller scale and less successful in both regeneration and job-creation terms.
There are few, if any, jobs for the former dockers and male manual workers
in Canary Wharf, but secretarial and administrative white-collar jobs abound
(Church and Frost, 1992). Arguably, the Canary Wharf redevelopment helped
to guarantee the long-term success of London as a leading international financial centre by providing large amounts of much-needed new high-quality
office space. As Rhodes and Tyler argue:

> In the wider local economy, the net impact of LDDC activities is lower,
> but even so, the LDDC generated an additional 23,000 jobs in Central
> London by increasing the supply of high grade office accommodation
> which led to a more competitive financial centre.
>
> (Rhodes and Tyler, 1998, p. 33)

More contentiously, they argue that

> In spite of vociferous comments to the contrary, over the life of the
> Corporation, the LDDC generated substantial benefits specifically for
> local communities and residents. The amount of new social housing is
> higher than it would have been in the absence of the Corporation.
>
> (Rhodes and Tyler, 1998, p. 33)

The claim about the scale of new social housing is very debatable, but the
improvement in transport links and environmental improvements for local
residents is without question. More generally, there is little doubt that the
redevelopment of Docklands and Canary Wharf has generated winners and
losers and that most of the winners are not locals, but this does not necessarily
undermine the overall success of the scheme. Its success needs to be measured
in terms of the massive redevelopment of Docklands in a way which was
almost unimaginable fifteen years ago. The area has been transformed from
derelict wharves, warehouses, factories and blighted landscapes to the biggest
financial office development in western Europe. It is extremely doubtful that
any of the alternative schemes being considered in the 1980s would have
been anywhere near as successful, particularly as the DJC's and LDSP's
plans for new factories and manufacturing jobs were already far behind
target and that far more manufacturing jobs were being lost than created.
To have gone with a manufacturing-based plan for Docklands regeneration
would have been to try to stem a tide that was already fast going out. In
retrospect, the LDDC vision appears to have been far more in tune with the

emergent new realities of employment growth in the post-industrial era. If the Docklands redevelopment has an Achilles' heel, it is not the stress on financial services employment but its willingness to allow increasing amounts of private sector luxury housing development with little attention being given to the need for more affordable housing in the area and within London as a whole. Clearly, the private sector will not provide this kind of housing of its own volition because it is unprofitable to do so, but arguably they could have been forced to include more socially rented housing under section 6 planning agreements in return for the grant of planning permission. This would have been very difficult, if not impossible, in the first half of the 1990s when the private speculative housing market in Docklands virtually collapsed, with a number of developers such as Kentish Homes going bankrupt, but it would have been possible in the boom years of the second half of the 1990s when developers were almost falling over themselves to acquire sites. This is not to argue that Docklands should have been developed with 50 per cent of more social housing, as the DJC proposed, as Tower Hamlets in 1981 already had 85 per cent of its households in the council sector. But more affordable housing could and should be provided to ensure that the local residents and their children can remain in the area. They should not be reduced to the status of children with their noses pressed against the glass of the new sweet shop but unable to go in and buy.

Other large, planned office developments

Canary Wharf will not be the last of major office developments in London, although it will probably be the biggest. The redevelopment of King's Cross has been scheduled since the late 1980s (Edwards, 1992), and although there has been some conversion of old industrial and warehouse property adjacent to the King's Cross canal basin into luxury apartments, the great bulk of the 40-ha site remains derelict and undeveloped as it has for most of the last thirty years, despite various development proposals. It seems that development is contingent on the construction of the new Eurostar rail link from Dover, but when this finally occurs there is no doubt that major development will take place.

Applications have gone in for a massive development in and around Paddington Basin in West London just west of Westway. The area comprises the old Paddington railway station goods yard and the Paddington canal basin, and miscellaneous other property. Unfortunately, because the site is in multiple ownership, with multiple developers the scheme is complex but it includes two towers, one of 42 storeys, the other of 43, plus low-rise offices on the site of the goods yard. Developers have pooled resources into the Paddington Regeneration Partnership. It is planned that 25,000 people will work here, 4,000 will live here and St Mary's hospital to double in size. Pearman comments that:

there were some hideous, if smaller, developments planned here in the 1980s and early 1990s. Luckily, the last recession killed most of them off. Today, much better architects are involved. But *one thing never changes: the relentless drive by developers to squeeze the maximum possible floor space out of their investment.* As the value of land in Paddington's golden triangle soars, the pressure is on to build more and more.

(Pearman, 2000; emphasis added)

In this respect, Pearman has hit the nail on the head. With expensive centrally located land and a market economy, the emphasis is constantly on maximising development potential, and hence the profitability, of sites subject to planning and design controls. If the financial and business services sector continues to grow, London, like all other major cities, will need more office space. The current market downturn is likely to prove short-lived and by the middle of the decade the pressure for land development could pick up.

New cultural landscapes

The redevelopment of parts of central and Inner London for offices is only part of the transformation of the centre. There has also been a substantial expansion of what can be termed 'new cultural landscapes' as part of the remaking of central London. The gradual conversion of Covent Garden from the old fruit and vegetable market into one of the most popular retailing and tourist attractions in London was culminated with the rebuilding of the Opera House, complete with shopping colonnades, in 2000. So too, transformation of St Pancras goods yard into the new British Library was completed in 1996, and in 2000 the new Tate Modern emerged from the shell of what had been Bankside power station. This has been linked with St Paul's via a spectacular new footbridge across the Thames. The whole South Bank is now undergoing a substantial transformation into a new cultural complex, with linkage of the Royal Festival Hall, the National Theatre, the Globe Theatre and Tate Modern via a riverside walkway. This extends eastwards to Tower Bridge and the Design Museum. In addition, the British Museum has completed the redevelopment of the main court which opens a large new area to the public, and Tate Britain on Millbank has recently opened seven new galleries (Newman and Smith, 2000; Teedon, 2001).

Outside the central area, a number of inner areas are in the process of redevelopment or rejuvenation into fashionable new 'cultural quarters'. Large parts of Clerkenwell, on both sides of St John Street, north of Smithfield market, are in the process of being made over into clubs, bars and restaurants, and the area around Brick Lane in Aldgate is similarly being converted from a Bangladeshi immigrant area into a centre for artistic production and consumption with new galleries, workshops, restaurants, bars and the like. A similar process is taking place in Hoxton (Attfield, 1998) and several fashionable Brit-Art galleries have opened in Hackney, just off the Kingsland Road, including the pioneering Flowers East and more recently Jay Jopling's White Cube Gallery. The most striking thing about these developments is that ten years ago they were almost unimaginable. The thought of anyone opening an art gallery in the East End (apart from the pioneering and long-established Whitechapel Art Gallery) would have provoked hoots of laughter. But, in a quite fundamental way, the artistic centre of gravity of London has shifted eastwards. The inner East End is now where much of the cutting edge is established. It is the equivalent of Montmartre in Paris at the end of the nineteenth century. Areas like Notting Hill have lost most of their cutting edge and have become upper-middle-class residential areas. The cultural action is now in the East End, not the West End, in the area covered by the well-known English children's rhyme 'Oranges and Lemons'. This was first published in 1744 as 'London Bells' and discovery of an early version has added another six, hitherto unknown, lines.

Two sticks and Apple,
Ring ye Bells at Whitechapple,
Old Father Bald Pate,
Ring ye Bells at Aldgate,
Maids in White Aprons,
Ring ye Bells at St Catherines,
Oranges and Lemmons,
Ring ye Bells at St Clemens,
When will you pay me,
Ring ye Bells at Old Bailey,
When I am rich,
Ring ye Bells at Fleetditch [now Shoreditch],
When will that be,
Ring ye Bells at Stepney,
When I am old,
Ring ye Great Bell at Pauls.
            (From Tommy Thumb's
            Pretty Song Book, 1744)

The areas referred to in 'Oranges and Lemons' have long been the cultural heartland of the East End, but it is symptomatic of recent developments that the renovation for sale as private apartments of Keeling House, a 1950s high-rise council block in Bethnal Green in the East End, designed by Denys Lasdun, a well-known modernist architect, was accompanied by a marketing brochure headlined 'Things are Looking Up', which proudly proclaimed:

Multi-racial, multi-cultural. Diverse talents. Creative influences. Chinese supermarkets. Indian restaurants, artists' studios . . . In the capital battle between East versus West, the style shapers and dictators are all facing in the same direction – East.

Although the marketing hype is visible, in terms of the changing class composition of London and its housing market, the council tenants are being gradually squeezed out of the new East End. The prices of the former council flats, now marketed as having spectacular and commanding views of the City, are such that most council tenants could not afford to live there even if they wanted to. The redevelopment of blocks such as this is indicative of the growing transformation of parts of the former East End of London into a desirable residential area to accompany the remaking of its former industrial landscape. Such areas possess one key attribute: close proximity to the City of London. Areas that were regarded 10–15 years ago as irredeemably downmarket and depressed are now within walking distance of the City, and in terms of the spatial restructuring of the post-industrial city they have been strongly revalorised. To this extent, the traditional division of London between the affluent middle- and upper-class residential West End and a poor working-class industrial East End is beginning to break down if only at the margins. In the new post-industrial city, proximity to the centre, and to the river, is increasingly important. As the industrial and transport land uses disappear to be replaced by offices, so the negative environmental factors which marked the area during the era of nineteenth- and twentieth-century industrialism are also disappearing. As a result the East End is slowly becoming socially and physically transformed. An emblematic symbol of the transformation is the conversion of the old Jewish soup kitchen for the poor in Spitalfields into luxury apartments. The high-rise council blocks of Tower Hamlets and Newham retain concentrations of the poor, but they are slowly being encircled with luxury riverside developments and conversions which are increasing the divisions between rich and poor in the new London.

Bibliography

Abu-Lughod, J. (1994) *From Urban Village to East Village: the Battle for New York's Lower East Side*, Oxford, Blackwell.

Adams, R. (2001) Don't worry mate, it's on expenses, *Guardian*, 10 July.

Adonis, A. and Pollard, S. (1997) *A Class Act: the Myth of Britain's Classless Society*, London, Hamish Hamilton.

Akbar, A. (2001) Middle class settlers fail to join community, *Independent*, 16 November.

Allen, J., Massey, D. and Cochrane, A. (1998) *Rethinking the Region*, London, Routledge.

Ambrose, P. and Colenutt, B. (1969) *The Property Machine*, Harmondsworth, Penguin.

Anderson, H. and Flatley, J. (1997) Contrasting London incomes: a social and spatial analysis, London, The Greater London Authority.

Angelo, L. (1995) Marching to its own drummer: why wage trends in New York City have diverged from the nation, *Urban Affairs Review*, 31, 1, 104–19.

Ascherson, N. (1986) London's new class: the great cash in, *Observer*, 25 May.

Ashworth, J. (2001) UK accountants top the global pay league, *The Times*, 8 December.

Atkinson, A.B. (1995) *Incomes and the Welfare State: Essays on Britain and Europe*, Cambridge, Cambridge University Press.

Atkinson, R. (2000a) Measuring gentrification and displacement in Greater London, *Urban Studies*, 37, 1, 149–66.

Atkinson, R. (2000b) Professionalisation and displacement in Greater London, *Area*, 23, 3, 287–96.

Atkinson, R. (2000c) Combating social exclusion in Europe: the new urban policy challenge, *Urban Studies*, 37, 5/6, 1037–55.

Attfield, A. (1998) The making of Hoxton's cultural quarter and its impact on urban regeneration in Hackney, *Rising East*, 1, 3, 133–55.

Auger, P. (2000) *The Death of Gentlemanly Capitalism: the Rise and Fall of London's Investment Banks*, Harmondsworth, Penguin.

August, O. (1997) Don't mention the euro as Germany prepares for E-day to dawn in the City, *The Times*, 15 July.

Ayers, C. (2001) Wall Street's big spenders call a halt to extravagance, *The Times*, 17 March.

Ayers, C. and Merrell, C. (2002) Worst job cuts on Wall Street for 25 years, *The Times*, 4 April.

Badcock, B. (1995) Notwithstanding the exaggerated claims, residential revitalization is changing the shape of some western cities: a response to Bourne, *Urban Studies*, 30, 191, 195.

Bailey, T. and Waldinger, R. (1991) The changing ethnic/racial division of labor, in J. Mollenkopf and M. Castells (eds) *Dual City: Restructuring New York*, New York, Russell Sage Foundation.

Baker, L. (2000) City concern at plan to merge London and German bourses, *Independent*, 18 April.

Baldwin, T. (2000) Doomed Dome was bankrupt by second month, *The Times*, 10 November.

Ball, M. (1994) The 1980s property boom, *Environment and Planning A*, 26, 671–95.

Ball, M. and Sutherland, D. (2001) *An Economic History of London, 1800–1914*, London, Routledge.

Banfield, E.C. (1974) *The Unheavenly City Revisited*, New York, Little, Brown & Co.

Banks, R. and Scanlon, R. (2000) Major economic trends in the 1980s and 1990s: London, in *The London–New York Study*, City of London Corporation.

Barker, P. (2002) Beyond the Dome, there is little left of Eastern Promise, *Independent*, 29 May.

Barlow, J. (1989) Planning the London conversion boom: flat developers and planners in the London housing market, *The Planner*, 75, 1, 18–21.

Barlow, J. and Gann, D. (1993) *Offices into Flats*, York, Joseph Rowntree Foundation.

Barlow, J. and Gann, D. (1995) Flexible planning and flexible buildings: reusing redundant office space, *Journal of Urban Affairs*, 17, 3, 263–76.

Barras, R. (1983) The office development cycle in London, *Land Development Studies*, 1, 1, 35–50.

Baum, S. (1997) Sydney, Australia: a global city? Testing the social polarization thesis, *Urban Studies*, 34, 11, 1881–1901.

Baum, S. (1999) Social transformations in the global city, Singapore, *Urban Studies*, 36, 1095–1117.

Beard, M. (2000) Hollywood invasion of London continues with De Niro's £3.25m purchase, *Independent*, 12 December.

Beaumont, J. and Hamnett, C. (2002) *Spatial Dimensions of Urban Social Exclusion and Integration: the Case of London*, Urbex Research Report No. 15, Amsterdam Study Centre for the Metropolitan Environment, University of Amsterdam.

Beaumont, J., Hamnett, C., Lee, P., Murie, A. and Oosthuizen, R. (2000) *Urban Social Exclusion: Comparative Statistical Analysis at National, Metropolitan, Local and Neighbourhood Level: London and Birmingham*, Urbex No. 7, University of Amsterdam.

Beauregard, R. (1991) Capital restructuring and the new built environment of global cities, New York and Los Angeles, *International Journal of Urban and Regional Research*, 15, 1, 90–105.

Beauregard, R. (1993) *Voices of Decline: the Postwar Fate of US Cities*, Oxford, Blackwell.

Beaverstock, J. (1991) Skilled international migration: an analysis of the geography of international secondments within large accountacy firms, *Environment and Planning A*, 23, 1133–46.

Beaverstock, J. (1994) Re-thinking skilled international labour migration: world cities and banking organizations, *Geoforum*, 25, 323–38.

Beaverstock, J. and Smith, J. (1996) Lending jobs to global cities, skilled international labour migration, investment banking and the City of London, *Urban Studies*, 33, 8, 1377–94.

Beaverstock, J.V. and Taylor, P.J. (1999) A roster of world cities, *Cities*, 16, 6, 445–58.

Beaverstock, J.V., Smith, R.G. and Taylor, P.J. (1999) The long arm of the law: London's law firms in a globalising world economy, *Environment and Planning A*, 31, 1857–76.

Beaverstock, J., Smith, R.G. and Taylor, P.J. (2000) Geographies of globalisation: US law firms in world cities, *Urban Geography*, 21, 1, 95–100.

Bell, D. (1973) *The Coming of Post-Industrial Society*, New York, Basic Books.

Bennetto, J. (2001) Police attack media for ignoring black killings, *Independent*, 27 December.

Bennetto, J. (2002) Police chief warns London faces 'justice by geography', *Independent*, 25 April.

Bentham, G. (1986) Socio-tenurial polarisation in the United Kingdom, 1953–83, *Urban Studies*, 23, 1, 57–62.

Berry, B.L.J. (1985) Islands of renewal in seas of decay, in P. Peterson (ed.) *The New Urban Reality*, Washington, DC, Brookings Institution.

Binney, M. (2001) High anxiety on Livingstone's point of view, *The Times*, 8 May.

Black, I. (1999) Rebuilding the heart of Empire: bank headquarters in the City of London, *Art History*, 22, 593–618. (Reprinted in D. Arnold (ed.) (1999) *The Metropolis and its Image: Constructing Identities for London, c. 1750–1950*, Oxford, Blackwell.)

Black, I. (2000) Spaces of capital: bank office building in the City of London, 1830–1870, *Journal of Historical Geography*, 26, 351–75.

Blackhurst, C. (2001) It is called the Canaletto effect. And big business claims it is holding back London, *Independent*, 21 April.

Bluestone, D. (1992) The pushcart evil, in D. Ward and O. Zunz (eds) *The Landscape of Modernity*, New York, Russell Sage.

Boggan, S. (2000) I don't know about the rest of Britain, or London, but round here it's like the Bronx, *Independent*, 1 July.

Boland, V. and Luce, E. (1998) Paris bourse plans rival to London–Frankfurt link, *Financial Times*, 18 July.

Boland, V. and Pretzlik, C. (2001) How the LSE managed to let Liffe slip away, *Financial Times*, 3 November.

Bondi, L. (1991) Gender divisions and gentrification: a critique, *Transactions of Institute of British Geographers*, 16, 2, 190–98.

Booth, C. ([1887, 1889, 1901] 1982) *Life and Labour of the People in London*, London, Macmillan.

Bourne, L.S. (1993) The demise of gentrification: a commentary and prospective view, *Urban Studies*, 14, 1, 95–107.

Bowley, G. (1997) MG to move metal trading to London, *Financial Times*, 1 May.

Braid, M. (2000) On Ocean estate, no one except the drugs gangs goes out after sundown, *Independent*, 13 April.

Bramley, G. (1994) An affordability crisis in British housing: dimensions, causes and policy impacts, *Housing Studies*, 9, 1, 103–24.

Bramley, G., Doogan, K., Leather, P., Murie, A. and Watson, E. (1988) *Homelessness and the London Housing Market*, SAUS Occasional Paper 32, School for Advanced Urban Studies, University of Bristol.

Braverman, H. (1974) *Labor and Monopoly Capital*, New York, Monthly Review Press.

Brenner, R. (2000) The boom and the bubble, *New Left Review*, 6, Nov./Dec., 5–44.

Bridge, G. (1994) Gentrification, class and residence, *Environment and Planning D*, 12, 31–51.

Bridge, G. (1995) The space for class? On class analysis in the study of gentrification, *Transactions of the Institute of British Geographers*, 20, 2, 236–47.

Brierley, D. (1998) UK set for euro bonanza: London to dominate trading in new currency from 4th January, *Observer*, 20 December.

Brint, S. (1991) Upper professionals: a high command of commerce, culture, and civic regulation, in J. Mollenkopf and M. Castells (eds) *Dual City: Restructuring New York*, New York, Russell Sage Foundation.

Brown, K. (2002) Southern discomfort, *Financial Times*, 19 March.

Browne, A. (1999) Would you pay £1 million for this three-bedroom terraced house?, *Observer*, 5 September.

Brownhill, S. (1990) *Developing London's Docklands: Another Great Planning Disaster*, London, Paul Chapman Publishing.

Brownhill, S. (1998) From exclusion to partnership? The LDDC and community consultation and participation, *Rising East*, 2, 2, 42–72.

Brownhill, S., Sharp, C., Jones, and Merrett, S. (1990) *Housing London: Issues of Finance and Supply*, York, Joseph Rowntree Foundation.

Bruegel, I. (1996) Gendering the polarisation debate: a comment on Hamnett's 'Social polarisation, economic restructuring and welfare state regimes', *Urban Studies*, 33, 1431–9.

Bruegel, I. (2000) The restructuring of London's labour force: migration and shifting opportunities, 1971–91, *Area*, 32, 1, 79–90.

Buck, N. (1994) Social divisions and labour market change in London: national, urban and global factors, unpublished paper for the ESRC London seminar, 28 October.

Buck, N. (1996) Social and economic change in contemporary Britain: the emergence of an urban underclass, in E. Mingione (ed.) *Urban Poverty and the Underclass: a Reader*, Oxford, Blackwell.

Buck, N. (1997) *Social Divisions and Labour Market Change in London*, Working Paper of the ESRC Research Centre on Micro-Social Change No. 97-25, Colchester, University of Essex.

Buck, N. and Gordon, I. (2000) Turbulence and sedimentation in the labour markets of late twentieth century metropoles, in G. Bridge and S. Watson (eds) *A Companion to the City*, Oxford, Blackwell.

Buck, N., Drennan, M. and Newton, K. (1992) Dynamics of the metropolitan economy, in S.S. Fainstein, M. Harloe and I. Gordon (eds) *Divided Cities: New York and London in the Contemporary World*, Oxford, Blackwell.

Buck, N., Gordon, I. and Young, K. (eds) (1986) *The London Employment Problem*, Oxford, Clarendon Press.

Budd, L. and Whimpster, S. (eds) (1992) *Global Finance and Urban Living: a Study of Metropolitan Change*, London, Routledge.

Bugler, J. (1985) The changing face of London: a tale of two cities, *The Listener*, 11 April.

Bull, P. and Church, A. (1996) The London tourist complex, in C.M. Law (ed.) *Tourism in Major Cities*, London, Mansell.

Burgers, J. (1996) No polarisation in Dutch cities? Inequality in a corporatist country, *Urban Studies*, 33, 1, 99–105.

Butler, T. (1996) People like us: the gentrification of Hackney in the 1980s, in T. Butler and M. Rustin (eds) *Rising in the East: the Regeneration of East London*, London, Lawrence & Wishart.

Butler, T. (1997) *Gentrification and the Middle Classes*, Ashford, Ashgate.

Butler, T. (1999) The new urban intermediaries? The new middle class and the remaking of London, *Journal des Anthropologues*, 77–78, 83–97.

Butler, T. and Hamnett, C. (1994) Gentrification, class and gender, *Society and Space/Environment and Planning D*, 12, 477–93.

Butler, T. and Robson, G. (2001a) Coming to terms with London: middle class communities in a global city, *International Journal of Urban and Regional Research*, 25, 1, 70–86.

Butler, T. and Robson, G. (2001b) Social capital, gentrification and neighbourhood change in London: a comparison of three South London neighbourhoods, *Urban Studies*, 38, 12, 2145–62.

Butler, T. and Rustin, M. (eds) (1996) *Rising in the East: the Regeneration of East London*, London, Lawrence & Wishart.

Butler, T. and Savage, M. (eds) (1995) *Social Change and the Middle Classes*, London, UCL Press.

Cadogan, G. (1999) What is there left to convert?, *Financial Times*, 30 October.

Cameron, S. and Evans, A. (1973) The economy of an agglomeration, *Regional Studies*, 7, 47–55.

Carpenter, J. and Lees, L. (1995) Gentrification in New York, London and Paris, *International Journal of Urban and Regional Research*, 19, 2, 286–34.

Castells, M. (1989) *The Informational City*, Oxford, Blackwell.

Castles, F. and Miller, M.J. (1993) *The Age of Migration: International Population Movements in the Modern World*, London, Macmillan.

Caulfield, J. (1994) *City Form and Everyday Life: Toronto's Gentrification and Critical Social Practice*, University of Toronto Press.

Champion, A.G. (1994) International migration and demographic change in the developed world, *Urban Studies*, 31, 4/5, 653–77.

Church, A. (1988) Urban regeneration in Docklands: a five year policy review, *Environment and Planning C: Government and Policy*, 6, 187–208.

Church, A. and Frost, M. (1992) The employment focus of Canary Wharf and the Isle of Dogs: a labour market perspective, *The London Journal*, 17, 2, 135–54.

Church, A. and Frost, M. (1995) The Thames Gateway – an analysis of the emergence of a sub-regional regeneration initiative, *The Geographical Journal*, 161, 2, 199–209.

Church, A. and Frost, M. (1998) Trickle down or tickle out: job creation and work-travel impacts of Docklands regeneration, *Rising East*, 2, 2, 73–103.

Church, A. and Frost, M. (2002) Global city tourism and the labour market in London, in S. Fainstein, L. Hoffman and D. Judd (eds) *Cities and Visitors*, Oxford, Blackwell.

City of London (2000) *The London–New York Study*, City of London Corporation.

Clark, G. (2002) London in the European financial services industry: locational advantage and product complementarities, *Journal of Economic Geography*, 2, 2, 1–21.

Clark, G. and Wojcik, D. (2001) The City of London in the Asian crisis, *Journal of Economic Geography*, 1, 1, 107–30.

Clark, W.A.V. (1996) Scale effects of international migration to the United States, *Regional Studies*, 30, 589–600.

Clark, W.A.V. (1999a) Mass migration and local outcomes: is international migration to the United States creating a new urban underclass?, *Urban Studies*, 35, 3, 371–83.

Clark, W.A.V. (1999b) *The Californian Melting Pot*, Berkeley, University of California Press.

Clark, W.A.V. and McNicholas, M. (1996) Re-examining economic and social polarisation in a multi-ethnic metropolitan area: the case of Los Angeles, *Area*, 28, 56–63.

Clement, B. (2000) Partners plan £1bn renaissance for King's Cross, *Independent*, 4 April.

Coakley, J. (1992) London as an international financial centre, in L. Budd and S. Whimpster (eds) *Global Finance and Urban Living*, London, Routledge.

Coakley, J. and Harris, L. (1983) *The City of Capital: London's Role as a Financial Centre*, Oxford, Blackwell.

Cobain, I. (2000) Churches strive to rid Peckham of gun culture, *Independent*, 1 December.

Cohen, N. (1999) High-rise game pays off for Reichmann, *Financial Times*, 20 March.

Cohen, N. (2000) The remarkable fall and rise of Canary Wharf, *Financial Times*, 21 October.

Cohen, N. (2001) Development continues at Canary Wharf, *Financial Times*, 10 March.

Cohen, R. (1981) The new international division of labour: multinational corporations and the urban hierarchy, in M. Dear and A.J. Scott (eds) *Urbanization and Urban Planning in Capitalist Society*, London, Methuen.

Coleman, A. (1985) *Utopia on Trial: Vision and Reality in Planned Housing*, London, Hilary Shipman.

Congdon, P. (1995) Socio-economic structure and health in London, *Urban Studies*, 32, 3, 523–49.

Conway, J. (1984) *Capital Decay: an Analysis of London's Housing*, London, SHAC.

Coopers, Lybrand Deloitte (1991) *London: World City Moving into the 21st Century*, London, HMSO.

Cope, N. (2001) Directors' pay surges as shares lose value, *Independent*, 31 October.

Coppock, J.T. (1964) The industries of London, in J.T. Coppock and H. Prince (eds) *Greater London*, London, Faber.

Counsell, G. (1988) Depression develops over Docklands, *Independent*, 10 September.

Coupland, A. (1992a) Every job an office job, in A. Thornley (ed.) *The Crisis of London*, London, Routledge.

Coupland, A. (1992b) Docklands: dream or disaster, in A. Thornley (ed.) *The Crisis of London*, London, Routledge.

Cowlard, K. (1992) City futures, in L. Budd and S. Whimpster (eds) *Global Finance and Urban Living: a Study of Metropolitan Change*, London, Routledge.

Cox, R. and Watt, P. (2002) Globalization, polarization and the informal sector: the case of paid domestic workers in London, *Area*, 34, 1, 39–47.

Coyle, D. (1997) *The Weightless Economy*, Oxford, Capstone.

Crampton, G. and Evans, A. (1992) The economy of an agglomeration: the case of London, *Urban Studies*, 29, 361–83.

Crilley, D. (1990) The disorder of John Short's New Urban Order, *Transactions of the Institute of British Geographers*, 15, 2, 232–8.

Crilley, D., Bryce, C., Hall, R. and Ogden, P. (1991) *New Migrants in London's Docklands*, Research Report No. 5, Department of Geography, Queen Mary College.

Cross, M. (1992) Race and ethnicity, in A. Thornley (ed.) *The Crisis of London*, London, Routledge.

Cross, M. and Johnson, M. (2002) Mobility denied: Afro-Caribbean labour and the British economy, in M. Cross and H. Etzinger (eds) *Last Illusions: Caribbean Minorities in Britain and the Netherlands*, London, Routledge, 73–105.

Cross, M. and Waldinger, R. (1992) Migrants, minorities and the ethnic division of labour, in S. Fainstein, M. Harloe and I. Gordon (eds) *Divided Cities: New York and London in the Contemporary World*, Oxford, Blackwell.

Curtis, S. (2001) Health in London, *Area*, 33, 1, 84–91.

Dale, A. and Bamford, A. (1989) Social polarisation in Britain, 1973–1982 – evidence from the General Household Survey: a comment on Pahl's hypothesis, *International Journal of Urban and Regional Research*, 13, 481–500.

Daley, P. (1998) Black Africans in Great Britain: spatial concentration and segregation, *Urban Studies*, 35, 10, 1703–25.

Daly, M. (1971) Characteristics of 12 clusters of wards in Greater London, *Greater London Council Department of Planning and Transportation, Research Report No. 13*, GLC.

Daniels, P.W. (1986) Foreign banks and metropolitan development: a comparison of London and New York, *Tijdschrift Economische en Sociale Geografie*, 77, 4, 269–87.

Daniels, P.W. (1991) *Services and Metropolitan Development: International Perspectives*, London, Routledge.

Daniels, P.W. (2000) Business and professional services in London, in *The London–New York Study*, City of London Corporation.

Daniels, P.W. and Bobe, J.M. (1993) Extending the boundary of the City of London? The development of Canary Wharf, *Environment and Planning A*, 25, 539–52.

Danson, M.W., Lever, W. and Malcolm, J.F. (1980) The inner city employment problem in Great Britain, 1952–76: a shift-share approach, *Urban Studies*, 17, 2, 193–210.

Davey, J. (2002a) Developers play for high stakes in London, *The Times*, 15 June.

Davey, J. (2002b) Tallest tower will bolster City's dominance, *The Times*, 21 June.

Davey, J. (2002c) Heron Tower joins Erotic Gherkin, *The Times*, 23 July.

Davis, E.P. and Latter, A.R. (1989) London as an international financial centre, *Bank of England Quarterly Bulletin*, 24, 4, 516–28.

Deakin, N. (1974) New commonwealth minorities in London, some issues, GLC RM draft.

Deakin, N. and Ungerson, C. (1977) *Leaving London: Planned Mobility and the Inner City*, London, Heinemann.

Dennis, R. (1978) The decline of manufacturing employment in Greater London, 1966–74, *Urban Studies*, 15, 63–74.

Department for Education and Science (DES) (1999) GCSE/GNVQ results for young people is England, 1999, LEA.

Department for Education and Science (DfES) (2003) *Youth Cohort Study: Activities and Experiences of 16 year olds, England and Wales*, DfES.

Department of the Environment (1976) *Lambeth Inner Area: Labour Market Study*, London.

Department of the Environment (1985) *Report of the Committee of Inquiry on the Management of Privately Owned Blocks of Flats*, London, DoE.

Department of Environment, Transport and the Regions (DETR) (1998) The Index of Local Deprivation, DETR.

Deskins, D. (1996) Economic restructuring, job opportunities and black social dislocation in Detroit, in J. O'Loughlin and J. Friedrichs (eds) *Social Polarisation in Post-Industrial Metropolises*, Berlin and New York, Walter de Gruyter.

Diamond, I. and Clarke, S. (1989) Demographic patterns among Britain's ethnic groups, in H. Joshi (ed.) *The Changing Population of Britain*, Oxford, Blackwell.

Dobbie, C. (1992) Fighting London's corner, *Independent*, 8 July.

Dobson, J. and McLaughlin, G. (2001) International migration to and from the United Kingdom, 1975–1999: consistency, change and implications for the labour market, *Population Trends*, 106, Winter, 29–38.

Donnison, D. (1967) *The Government of Housing*, Harmondsworth, Penguin.

Donnison, D. and Eversley, D. (1973) *London: Urban Patterns, Problems and Policies*, London, Heinemann.

Doran, J. and Ayers, C. (2002) Partners to scoop $1m each in Price Waterhouse Coopers float, *The Times*, 30 April.

Dorling, D. *et al.* (2000) The ghost of Christmas past: health effects of poverty in London in 1896 and 1991, *British Medical Journal*, vol. 321, 1547–51.

Drennan, M. (1991) The decline and rise of the New York economy, in J. Mollenkopf and M. Castells (eds) *Dual City: Restructuring New York*, New York, Russell Sage Foundation.

Drennan, M.P. (1999) The dominance of international finance by London, New York and Tokyo, in P. Daniels and W.F. Lever (eds) *The Global Economy in Transition*, Harlow, Longman.

Drew, D., Gray, J. and Sporton, D. (1997) Ethnic differences in the educational participation of 16–19 year olds, in V. Karn (ed.) *Ethnicity in the 1991 Census*, Vol. IV, London, HMSO.

Dunleavy, P. (1981) *The Politics of Mass Housing in Britain, 1945–75*, Oxford, Clarendon Press.

Dunn, R., Forrest, R. and Murie, A. (1987) The geography of council house sales in England, 1979–85, *Urban Studies*, 24, 1, 47–59.

Dunning, J.H. and Morgan, E.V. (1971) *An Economic Study of the City of London*, London, George Allen & Unwin.

Durman, P. (1998) Paris to join Europe Alliance, *Independent*, 20 November.

Dyos, H.J. (1967) The slums of Victorian London, *Victorian Studies*, September, 10, 1, 5–40.

Dyos, H.J. (1968) The speculative builders and developers of Victorian London, *Victorian Studies*, 11, 641–90.

Eade, J. *et al.* (1996) The Bangladeshis: the encapsulated community, in C. Peach (ed.) *Ethnicity in the 1991 Census*: Vol. 2, *The Ethnic Minority Populations of Great Britain*, London, HMSO.

Eade, J. and Mele, C. (1998) Global processes and customised landscapes: the 'eastern promise' of New York and London, *Rising East*, 1, 3, 52–73.

The Economist (2000) The geography of cool, *The Economist*, 15 April.

The Economist (2001) Rebuilding London: two versions of market economics are at war over the redevelopment of Spitalfields, *The Economist*, 21 April.

The Economist (2002) London: bright lights, big city, *The Economist*, 16 March.

Edwards, M. (1992) A microcosm: redevelopment proposals at King's Cross, in A. Thornley (ed.) *The Crisis of London*, London, Routledge.

Ehrenreich, J. and Ehrenreich, B. (1979) The professional managerial class, *Radical America*, 11, 7–31.

Eisenhammer, J. (1995a) EU regulation poses biggest threat to the City, *Independent*, 13 March.

Eisenhammer, J. (1995b) Un-British Square Mile still leads the finance league, *Independent*, 13 March.

Eisenhammer, J. (1996) Square Mile under threat from Amsterdam, *Independent*, 9 January.

Elkin, S. (1974) *Politics and Land Use Planning: the London Experience*, Cambridge, Cambridge University Press.

Engels, F. (1969) *The Condition of the Working Class in England*, London, Panther.

Esping-Andersen, G. (1993) *Changing Classes: Stratification and Mobility in Post-industrial Societies*, London, Sage.

Esteban, J.M. and Ray, D. (1994) On the measurement of polarisation, *Econometrica*, 62, 819–51.

Evans, A.W. (1973) The location of the headquarters of UK industrial companies, *Urban Studies*, 10, 387–95.

Eversley, D. (1972) Rising costs and static incomes: some economic consequences of regional planning in London, *Urban Studies*, 9, 347–68.

Eversley, D. (1984) Does London need strategic planning?, *The London Journal*, 10, 1, 13–45.

Eversley, D. (1992) Urban disadvantage and racial minorities in the UK, in M. Cross (ed.) *Ethnic Minorities and Industrial Change in Europe and North America*, Cambridge, Cambridge University Press.

Fainstein, N. and Fainstein, S. (1987) Economic restructuring and the politics of land use planning in New York City, *American Planning Association Journal*, Spring, 237–48.

Fainstein, S. (1990) Economics, politics and development policy: the convergence of New York and London, *International Journal of Urban and Regional Research*, 14, 4, 553–75.

Fainstein, S. (1994a) Government programs for commercial redevelopment in poor neighbourhoods: the case of Spitalfields, East London and downtown Brooklyn, *Environment and Planning A*, 26, 215–34.

Fainstein, S. (1994b) *The City Builders: Property, Politics and Planning in London and New York*, Oxford, Blackwell.

Fainstein, S. and Harloe, M. (2000) Ups and downs in the global city: London and New York at the millennium, in G. Bridge and S. Watson (eds) *A Companion to the City*, Oxford, Blackwell.

Fainstein, S., Harloe, M. and Gordon, I. (eds) (1992) *Divided Cities: New York and London in the Contemporary World*, Oxford, Blackwell.

Feagin, J. (1987) The secondary circuit of capital: office construction in Houston, Texas, *International Journal of Urban and Regional Research*, 11, 172–92.

Feagin, J.R. and Smith, M.P. (1989) Cities in the new international division of labour: an overview, in M.P. Smith and J.R. Feagin (eds) *The Capitalist City*, Oxford, Blackwell.

Feigenbaum, H., Henig, J. and Hamnett, C. (1998) *Shrinking the State: the Political Underpinnings of Privatisation*, Cambridge, Cambridge University Press.

Fieldhouse, E.A. (1999) Ethnic minority unemployment and spatial mismatch: the case of London, *Urban Studies*, 36, 9, 1569–96.

Fielding, A. (1989) Inter-regional migration and social change: a study of south-east England, *Transactions of the Institute of British Geographers*, NS, 14, 24–36.

Fielding, A. (1991) Migration and social mobility: South East England as an escalator region, *Regional Studies*, 26, 1, 1–15.

Fielding, A. (1993) Mass migration and economic restructuring, in R. King (ed.) *Mass Migration in Europe: the Legacy and the Future*, London, Belhaven.

Fielding, A. (1995) Migration and middle class formation in England and Wales, in T. Butler and M. Savage (eds) *Social Change and the Middle Classes*, London, UCL Press.

Fielding, N. (2001) Barclays bankers who blew £44,000 on dinner, *Sunday Times*, 15 July.

Fitch, R. (1993) *The Assassination of New York*, London, Verso.

Fitch, R. (1994) Explaining New York's aberrant economy, *New Left Review*, 207, 17–48.

Florio, S. and Brownhill, S. (2000) Whatever happened to criticism? Interpreting the London Docklands Development Corporation's obituary, *Cities*, 4, 1, 53–64.

Ford, R. (2001) Foreigners' points win prize of skilled job in Britain, *The Times*, 14 December.

Ford, R. (2002) Migration record raises population by 183,000 a year, *The Times*, 7 May.

Ford, T. and Champion, T. (2000) Who moves into, out of and within London? An analysis of the 1991 Census 2% sample of anonymised records, *Area*, 32, 3, 259–70.

Forman, C. (1989) *Spitalfields: a Battle for Land*, London, Hilary Shipman.

Forrest, R. and Murie, A. (1988) *Selling the Welfare State: the Privatisation of Public Housing*, London, Routledge.

Forrest, R. and Murie, A. (1990) *Residualisation and Council Housing: a Statistical Update*, School for Advanced Urban Studies, Working Paper 91, Bristol, SAUS.

Forrest, R. and Murie, A. (1995) *Housing and Family Wealth: Comparative International Perspectives*, London, Routledge.

Foster, J. (1992) Living with the Docklands redevelopment: the community view from the Isle of Dogs, *The London Journal*, 17, 2, 170–83.

Foster, J. (1999) *Docklands: Cultures in Conflict, Worlds in Collision*, London, UCL Press.

Fothergill, S., Gudgin, G., Kitson, M. and Monk, S. (1986) Deindustrialization of the city, in R. Martin and B. Rowthorn (eds) *The Geography of Deindustrialization*, London, Macmillan.

Frank, R.H. and Cook, P.J. (1995) *The Winner-Take-All Society: Why the Few at the Top Get So Much More Than the Rest of Us*, Harmondsworth, Penguin.

Frayn, M. (1967) *Towards the End of the Morning*, London, Faber & Faber.

Frean, A. (2001a) Boom in British ethnic minority population, *The Times*, 21 September.

Frean, A. (2001b) Middle class immigrants to boost South East, *The Times*, 14 December.

Friedmann, J. (1986) The world city hypothesis, *Development and Change*, 17, 1, 69–84.

Friedmann, J. (1995) Where we stand: a decade of world city research, in P.L. Knox and P.J. Taylor (eds) *World Cities in a World System*, Cambridge, Cambridge University Press.

Friedmann, J. and Wolff, K. (1982) World city formation: an agenda for research and action, *International Journal of Urban and Regional Research*, 6, 3, 309–44.

Frost, M. and Church, A. (2001) Global city tourism and the labour market in London, in S. Fainstein, L. Hoffman and D. Judd (eds) *Cities and Visitors*, Oxford, Blackwell.

Frost, M. and Spence, N. (1991a) Employment changes in Central London in the 1980s. I: The record of the 1980s, *The Geographical Journal*, 157, 1, 1–12.

Frost, M. and Spence, N. (1991b) Employment changes in Central London in the 1980s. II: Understanding recent forces for change and future development constraints, *The Geographical Journal*, 157, 2, 125–35.

Frost, M. and Spence, N. (1993) Global city characteristics and Central London's employment, *Urban Studies*, 30, 3, 547–58.

Gale, D.E. (1978) Middle class resettlement in older urban neighbourhoods: the evidence and the implications, *American Planning Association Journal*, July, 293–304.

Gapper, J. and Denton, N. (1996) *All That Glitters: the Fall of Barings*, London, Hamish Hamilton.

Garfield, A. (2000a) German's hours deals blow to City, *Independent*, 11 April.

Garfield, A. (2000b) London and Frankfurt tie the knot in first wave of mergers, *Independent*, 4 May.

Garfield, A. (2000c) Frankfurt deal in disarray after OM launches £822m hostile bid for LSE, *Independent*, 30 August.

Garner, R. (2002) Middle classes abandoning state schools, *Independent*, 8 June.

Garner, R. and Pyke, N. (2002) Stop blaming schools, black head tells parents, *Independent*, 14 March.

Garside, P.L. (1983) Intergovernmental relations and housing policy in London 1919–70: the density and location of council housing, *The London Journal*, 9, 1, 39–57.

Gibb, F. (2001a) 'Corrrupt' Dame Shirley ordered to pay £26.4m, *The Times*, 14 December.

Gibb, F. (2001b) Top barristers push earnings to £500,000, *The Times*, 3 September.

Glancy, J. (1999) The other side of Shoreditch, *Guardian, Space*, 18 November.

Glass, R. (1963) Introduction to *London: Aspects of Change*, London, Centre for Urban Studies. (Reprinted in R. Glass (ed.) (1989) *Clichés of Urban Doom*, Oxford, Blackwell.)

Glass, R. (1968) Urban sociology in Great Britain, in R. Pahl (ed.) *Readings in Urban Sociology*, Oxford, Pergamon Press.

Glass, R. (1973) The mood of London, in D. Donnison and D. Eversley (eds) *London: Urban Patterns, Problems and Policies*, London, Heinemann. (Reprinted in R. Glass (ed.) (1989) *Clichés of Urban Doom*, Oxford, Blackwell.)

Glendenning, M. and Muthesias, S. (1994) *Tower Block: High-Rise Housing in England, Wales, Scotland and Northern Ireland*, New Haven, Conn., Yale University Press.

Goddard, J.B. and Smith, I.J. (1978) Changes in the corporate control in the British urban system, 1972–77, *Environment and Planning A*, 10, 1073–84.

Goodheart, C.A.E. (1987) The economics of 'Big Bang', *Midland Bank Review*, summer.

Goodwin, M. (1991) Replacing a surplus population: the policies of the London Docklands Development Corporation, in J. Allen and C. Hamnett (eds) *Housing and Labour Markets*, London, Unwin Hyman.

Goodwin, M. (1996) Governing the spaces of difference: regulation and globalisation in London, *Urban Studies*, 33, 8, 1395–1406.

Gordon, I. (1996) Family structure, educational achievement and the inner city, *Urban Studies*, 33, 3, 407–24.

Gordon, I. (1999) Move on up the car: dealing with structural unemployment in London, *Local Economy*, May.

Gordon, I. (2000) Demography, labour force and income, in *the London–New York Study*, City of London Corporation.

Gordon, I. (2001) Unpacking 'competitiveness' as a governance issue for London, in S. Syrett and R. Baldock (eds) *Governing London: Competitiveness and Regeneration for a Global City*, London, Middlesex University Press.

Gordon, I. and Harloe, M. (1991) A dual to New York? London in the 1980s, in J. Mollenkopf and M. Castells (eds) *Dual City: Restructuring New York*, New York, Russell Sage Foundation.

Gould, W. and Findlay, A. (1994) *Population Migration and the Changing World Order*, Chichester, Wiley & Sons.

Grabher, G. (2001) Ecologies of creativity: the village, the group and the hierarchic organization of the British advertising industry, *Environment and Planning A*, 33, 351–74.

Graham, D. and Spence, N. (1995) Contemporary deindustrialisation and tertiarisation in the London economy, *Urban Studies*, 32, 6, 885–911.

Graham, S. (1999) Global grids of glass: on global cities, telecommunication and planetary urban networks, *Urban Studies*, 36, 5–6.

Gray, C. (2001) A flaming cross marks gang's hold on estate, *Independent*, 12 June.

Gray, C. (2002) Capital's tourist chief quits over visitor slump that 'may rob London of world city status', *Independent*, 12 April.

Greater London Authority (GLA) (2000) *Homes for a World City*, London, Lord Mayor's Housing Commission.

Greater London Authority (2001) *Towards the London Plan: Initial Proposals for the Mayor's Spatial Development Strategy*, London, GLA (May).

Greater London Authority (2002a) *Planning for London's Growth*, GLA Spatial Development Strategy Technical Document No. 3.

Greater London Authority (2002b) *Creativity: London's Core Business*, London, GLA.

Greater London Council (GLC) (1969) *Greater London Development Plan*, London, GLC.

Greater London Council (1976) *Home Sweet Home: Housing Designed by the London County Council and GLC Architects, 1880–75*, London, GLC.

Greater London Council (1985a) *The State of the Art or the Art of the State: Strategies for the Cultural Industries in London*, London, GLC.

Greater London Council (1985b) *The London Industrial Strategy*, London, GLC.

Green, A.E. and Owen, D. (1995) Ethnic minority groups in regional and local labour markets in Britain: a review of data sources and associated issues, *Regional Studies*, 29, 8, 729–35.

Green, D. (1986) A map for Mayhew's London: the geography of poverty in the mid-nineteenth century, *The London Journal*, 11, 2, 115–26.

Green, D. (1995) *From Artisans to Paupers: Economic Change and Poverty in London, 1790–1870*, Aldershot, Scolar Press.

Griapos, P. (1977) The closure of firms in the inner city: the southeast London case, 1970–75, *Regional Studies*, 11, 1–6.

Griffiths, C. (2000) Legal eagles swoop to grab the power and the money, *The Times*, 7 October.

Haddon, R. (1970) A minority in a welfare state society: location of West Indians in the London housing market, *New Atlantis*, 2, 80–113.

Hall, P. (1962) *The Industries of London since 1861*, London, Hutchinson.

Hall, P. (1964) Industrial London: a general view, in J.T. Coppock and H. Prince (eds) *Greater London*, London, Faber.

Hall, P. (1966) *The World Cities*, London, Weidenfeld & Nicolson.

Hall, P. (1989) *London 2001*, London, Unwin Hyman.

Hall, P. (1992) Cities in the informational economy, *Urban Futures*, Special Issue 5, 13–26.

Hall, P. (2000) Creative cities and economic development, *Urban Studies*, 37, 4, 639–51.

Hall, R. and Hall, J. (1995) Missing in the 1991 census: a million persons – and £ millions in grants?, *Area*, 27, 1, 53–61.

Hall, R. and Ogden, P. (1992) The social structure of new migrants to London Docklands: recent evidence from Wapping, *The London Journal*, 17, 2, 153–69.

Halsey, A.H. (1988) *British Social Trends Since 1900* (2nd edn), Basingstoke, Macmillan.

Halsey, A.H. (2000) A hundred years of social change, *Social Trends*, 30, 15–20, ONS.

Halsey, A.H., Heath, A.F. and Ridge, J.M. (1980) *Origins and Destinations: Family, Class and Education in Modern Britain*, Oxford, Clarendon Press.

Hamilton, A. (2000) As bonuses go, £1m makes a poor Christmas, *The Times*, 18 December.

Hamilton, K. (2001a) City fights back with giant tower, *Sunday Times*, 9 September.

Hamilton, K. (2001b) Spurned Stock Exchange seeks new bride, *Sunday Times*, 4 November.

Hamnett, C. (1973) Improvement grants as an indicator of gentrification in London, *Area*, 5, 4, 252–61.

Hamnett, C. (1976) Social change and social segregation in Inner London, 1961–71, *Urban Studies*, 13, 261–71.

Hamnett, C. (1984) Gentrification and residential location theory: a review and assessment, in D.T. Herbert and R.J. Johnston (eds) *Geography and the Urban Environment*, Chichester, Wiley & Sons.

Hamnett, C. (1986) The changing socio-economic structure of London and the South-East, 1961–81, *Regional Studies*, 20, 391–406.

Hamnett, C. (1987) A tale of two cities: socio-tenurial polarisation in London and the South East, 1966–1981, *Environment and Planning A*, 19, 537–56.

Hamnett, C. (1989a) Spatial and social segmentation of the London owner occupied housing market: an analysis of the flat conversion sector, in M. Breheny and P. Congdon (eds) *Growth and Change in a Core Region: the Case of South East England*, London, Pion.

Hamnett, C. (1989b) The political geography of housing in contemporary Britain, in J. Mohan (ed.) *The Political Geography of Contemporary Britain*, London, Macmillan.

Hamnett, C. (1990) Migration and residential social change: a longitudinal analysis of London, *Revue de Géografie de Lyons*, 65, 155–63.

Hamnett, C. (1991a) Labour markets, housing markets and social restructuring in a global city: the case of London, in J. Allen and C. Hamnett (eds) *Housing and Labour Markets*, London, Unwin Hyman.

Hamnett, C. (1991b) The relationship between residential migration and housing tenure in London, 1971–81: a longitudinal analysis, *Environment and Planning A*, 23, 1147–62.

Hamnett, C. (1991c) The blind men and the elephant: towards a theory of gentrification, *Transactions of the Institute of British Geographers*, 16, 173–89.

Hamnett, C. (1994a) Social polarisation in global cities: theory and evidence, *Urban Studies*, 31, 3, 401–24.

Hamnett, C. (1994b) Socio-economic change in London: professionalisation not polarisation, *Built Environment*, 20, 3, 192–203.

Hamnett, C. (1996) Social polarisation, economic restructuring and welfare state regimes, *Urban Studies*, 33, 8, 1407–30.

Hamnett, C. (1997) A stroke of the Chancellor's pen: the social and regional impact of the Conservatives' 1988 higher rate tax cuts, *Environment and Planning A*, 29, 129–47.

Hamnett, C. (1999) *Winners and Losers: the Home Ownership Market in Modern Britain*, London, Taylor & Francis.

Hamnett, C. (2000) Gentrification, postindustrialism, and industrial and occupational restructuring global cities, in G. Bridge and S. Watson (eds) *A Companion to the City*, Oxford, Blackwell.

Hamnett, C. (2001) Social segregation and social polarization, in R. Paddison (ed.) *Handbook of Urban Studies*, London, Sage.

Hamnett, C. and Cross, D. (1998a) Social polarisation and inequality in London: the earnings evidence, 1979–95, *Environment and Planning C*, 16, 659–80.

Hamnett, C. and Cross, D. (1998b) Social change, social polarisation and income inequality in London, 1979–1993, *Geojournal*, 46, 39–50.

Hamnett, C. and Randolph, W. (1982) How far will London's population fall?: a commentary on the 1981 census, *The London Journal*, 8, 1, 96–100.

Hamnett, C. and Randolph, W. (1983) The changing tenure structure of the Greater London housing market, *The London Journal*, 9, 2, 153–64.

Hamnett, C. and Randolph, W. (1984) The role of landlord disinvestment in housing market transformation: an analysis of the flat break-up market in central London, *Transactions of the Institute of British Geographers*, NS, 9, 3, 259–79.

Hamnett, C. and Randolph, W. (1986a) Tenurial transformation and the flat breakup market in London, in N. Smith and P. Williams (eds) *Gentrification of the City*, London, Allen & Unwin.

Hamnett, C. and Randolph, W. (1986b) Ethnic minorities in the London labour market: a longitudinal analysis, 1971–81, *New Community*, 14, 333–46.

Hamnett, C. and Randolph, W. (1987) The residualisation of council housing in Inner London, 1971–81, in D. Clapham and J. English (eds) *Public Housing: Current Trends and Future Developments*, Beckenham, Croom Helm.

Hamnett, C. and Randolph, W. (1988a) *Cities, Housing and Profits: Flat Break-up and the Decline of Private Renting*, London, Hutchinson.

Hamnett, C. and Randolph, W. (1988b) Housing and labour markets change in London: a longitudinal analysis, 1971–81, *Urban Studies*, 25, 380–98.

Hamnett, C. and Randolph, W. (1992) Racial minorities in the London labour and housing markets: a longitudinal analysis, 1971–81, in M. Cross (ed.) *Ethnic Minorities and Industrial Change in Europe and North America*, Cambridge, Cambridge University Press.

Hamnett, C. and Whitelegg, D. (2001) *The Loft Conversion Market in Central London: a Case Study of Clerkenwell*, Final Report to the Economic and Social Research Council.

Hamnett, C. and Williams, P. (1980) Social change in London: a study of gentrification, *The London Journal*, 6, 1, 51–66.

Harloe, M. (1992) Housing inequality and social structure in London, *Housing Studies*, 7, 189–204.

Harloe, M., Issacharoff, R. and Minns, R. (1973) The organisation of housing policy in Inner London: the Lambeth experience, in D. Donnison and D. Eversley (eds) *London: Urban Patterns, Problems and Policies*, London, Heinemann.

Harris, M. (1973) Aspects of social polarisation, in D. Donnison and D. Eversley (eds) *London: Urban Patterns, Problems and Policies*, London, Heinemann.

Harrison, B. and Bluestone, B. (1988) *The Great U-turn, Corporate Restructuring and the Polarising of America*, New York, Basic Books.

Harrison, C. (2000) A matter of Liffe and death, *Metro*, 24 November.

Harrison, P. (1983) *Inside the Inner City: Life under the Cutting Edge*, Harmondsworth, Penguin.

Harvey, D. (1973) *Social Justice and the City*, London, Edward Arnold.

Harvey, D. (1975) Class structure in a capitalist society and the theory of residential differentiation, in R. Peel, M. Chisholm and P. Haggett (eds) *Processes in Physical and Human Geography*, London, Heinemann.

Harvey, D. (1978) The urban process under capitalism: a framework for analysis, *International Journal of Urban and Regional Research*, 2, 2, 101–32.

Harvey, D. (1985) *The Urbanization of Capital*, Baltimore, Johns Hopkins University Press.

Harvey, D. (1995) Cities or urbanization, *City*, 1, 38–61.

Heath, A. and McMahon, D. (1997) Education and occupational attainments: the impact of ethnic origins, in V. Karn (ed.) *Ethnicity in the 1991 Census*, Vol. IV, London, HMSO.

Heath, T. and Hill, R. (2002) *Asylum Statistics: UK 2001*, Home Office Research Report.

Hebbert, M. (1992) One 'planning disaster' after another: London Docklands, 1970–1992, *The London Journal*, 17, 2, 115–34.

Hebbert, M. (1999) *London: More by Fortune, than Design*, Chichester, John Wiley & Sons.

Heilbrun, J. (1992) Art and culture as central place functions, *Urban Studies*, 29, 2, 205–16.

Henderson, J. and Karn, V. (1984) Race, class and the allocation of public housing in Britain, *Urban Studies*, 21, 115–28.

Henneberry, J. (1999) Convergence and divergence in regional office development cycles, *Urban Studies*, 36, 9, 1439–66.

Hills, J. (1995) *Inequality in Britain, Report of the Inquiry*, York, Joseph Rowntree Foundation.

Hills, J. (1996) *New Inequalities: the Changing Distribution of Income and Wealth in the United Kingdom*, Cambridge, Cambridge University Press.

Hills, J. (2002) *Understanding Social Exclusion*, Oxford, Oxford University Press.

Hirst, P. and Thompson, G. (1999) *Globalisation in Question* (2nd edn), Oxford, Polity Press.

Hobcraft, J. (2000) The role of schooling and educational qualifications in the emergence of adult social exclusion, *CASE Paper 43*, Centre for Analysis of Social Exclusion, LSE.

Hoggart, K. (1991) The changing world of corporate control centres, *Geography*, 76, 1, 109–20.

Hoggart, R. (1957) *The Uses of Literacy*, Harmondsworth, Penguin.

Holmans, A.E. (1987) *Housing Policy in Britain*, London, Croom Helm.

Home Office (2002) *Asylum Statistics United Kingdom 2001*, 31 July.

Housing and Public Health Committee (1937) *London Housing*, London County Council.

Howe, G. (1986) Does it matter where I live?, *Transactions of the Institute of British Geographers*, 387–414.

Hudson, R. and Williams, A.M. (1995) *Divided Britain* (2nd edn), Chichester, John Wiley & Sons.

Hughes, C. (2001a) More jobs to be cut as City faces massive shake-up, *Independent*, 19 October.

Hughes, C. (2001b) Cantor moves Frankfurt and Paris operations to London, *Independent*, 16 October.

Hughes, C. (2001c) Liffe rejects Stock Exchange offer in favour of Euronext, *Independent*, 30 October.

Hughes, C. (2001d) Schroders pay deal prompts City row, *Independent*, 16 November.

Hughes, C. (2002) Suitors line up for hand of LSE, *Independent*, 15 May.

Hughes, C. and Griffiths, K. (2001) Boardroom pay soars in FTSE 100 Companies, *Independent*, 17 December.

Hyde, W. and Dilnot, S. (2000) The financial services sector in London, in *The London–New York Study*, City of London Corporation.

Income Data Services (2001) *The Directors Pay Report 2000–01*, London, IDS.

Inland Revenue (2002) *Inland Revenue Statistics, 2001*, London, HMSO.

Jacobs, J.M. (1992) Edge of Empire, in L. Budd and S. Whimpster (eds) *Global Finance and Urban Living*, London, Routledge.

Jacobs, J.M. (1994) The battle of Bank Junction: the contested iconography of capital, in S. Corbridge, R. Martin and N. Thrift (eds) *Money, Power and Space*, Oxford, Blackwell.

Jackson, A. (1974) *Semi-detached London*, London, George Allen & Unwin.

Jackson, B. and Marsden, B. (1966) *Education and the Working Class*, Harmondsworth, Penguin.

Jager, M. (1986) Class definition and the aesthetics of gentrification: Victoriana in Melbourne, in N. Smith and N. Williams (eds) *Gentrification of the City*, London, Allen & Unwin.

Jameson, A. (2000a) Struggle to unearth the capital's best kept secret, *The Times*, 30 December.

Jameson, A. (2000b) Deutsche Bank may quit City for Canary Wharf, *The Times*, 26 August.

Jameson, A. (2000c) McGraw Hill takes space in Canary Wharf, *The Times*, 6 October.

Jameson, A. (2000d) Canary Wharf adds to land bank, *The Times*, 18 October.

Jameson, A. (2000e) London housebuilders see no profit in Livingstone's plan, *The Times*, 27 October.

Jameson, A. (2001a) Canary Wharf set for capital return, *The Times*, 10 March.

Jameson, A. (2001b) Out-of-town spectre looms over Canary Wharf, *The Times*, 20 July.

Jameson, A. (2001c) Skyscraper reaches for residential record, *The Times*, 13 December.

Jenkins, S. (1975) *Landlords to London: the Story of a Capital and its Growth*, London, Constable.

Jenkins, S. (1999) Thank God for the City's foreigners, *Evening Standard*, 4 February.

Jenkins, S. (2000) Why Frankfurt can never measure up to London, *Evening Standard*, 22 October.

Jenkins, S. (2001) Livingstone's passion for tall storeys, *The Times*, 25 February.

Jenkins, S.P. and Cowell, F.A. (1994) Dwarfs and giants in the 1980s: trends in the UK income distribution, *Fiscal Studies*, 15, 99–118.

Jennings, J.H. (1971) Geographical implications of the municipal housing programme in England and Wales, 1919–39, *Urban Studies*, 8, 121–38.

Jessop, B., Bonnett, K., Bromley, S. and Ling, T. (1988) *Thatcherism*, Cambridge, Polity.

Johnson, J.H. (1964) The suburban expansion of housing in London 1918–1939, in J.T. Coppock and H.C. Prince (eds) *Greater London*, London, Faber.

Johnston, R.J. (1979) The spatial impact of fiscal changes in Britain: regional policy in reverse, *Environment and Planning A*, 11, 1439–44.

Johnston, R.J., Pattie, C.J. and Allsop, J.G. (1988) *A Nation Dividing*, London, Longman.

Joseph, C. (2000) Just call it Silicon Lane, *Independent*, 20 November.

Judge, E. and Jameson, A. (2000) London homes must be for less well-off, *The Times*, 17 November.

Kaletsky, A. (1995) Banking on London's freedom, *The Times*, 16 June.

Kaletsky, A. (1999) EMU is not the real threat to the future of the City, *The Times*, 11 May.

Kamo, T. (2000) Aftermath of globalisation? East Asian economic turmoil and Japanese cities adrift, *Urban Studies*, 37, 12, 2145–66.

Karn, V. (1997) *Ethnicity in the 1991 Census*. Vol. IV: *Employment, Education and Housing Among the Ethnic Minority Populations of Britain*, ONS, London, HMSO.

Kennedy, R. (1991) *London: World City Moving into the 21st Century: a Research Project*, London, HMSO.

King, A.D. (1990) *Global Cities: Post-Imperialism and the Internationalization of London*, London, Routledge.

King, R. (1993) *Mass Migration in Europe: the Legacy and the Future*, London, Belhaven Press.

Kleinman, M. and Whitehead, C. (1987) Local variations in the sale of council houses in England, 1979–1984, *Regional Studies*, 21, 1, 1–12.

Kloosterman, R. (1996) Double Dutch: polarisation trends in Amsterdam and Rotterdam after 1980, *Regional Studies*, 30, 467–76.

Kloosterman, R.C. and Lamboy, J. (1992) The Randstad – a welfare region, in F. Dieleman and S. Musterd (eds) *The Randstad: a Research and Policy Laboratory*, Dordrecht, Kluwer.

Knox, P. (1991) The restless urban landscape: economic and sociocultural change and the transformation of Washington DC, *Annals Association of American Geographers*, 81, 181–209.

Knox, P. (1995) World cities in a world system, in P. Knox and P. Taylor (eds) *World Cities in a World System*, Cambridge, Cambridge University Press.

Knox, P. and Taylor, P. (1995) *World Cities in a World System*, Cambridge, Cambridge University Press.

Kumar, K. (1978) *Prophecy and Progress: the Sociology of Industrial and Post Industrial Society*, Harmondsworth, Penguin.

Kuttner, B. (1983) The declining middle, *Atlantic Monthly*, July, 60–72.

Kynaston, D. (1994) *The City of London*. Vol. I: *A World Of Its Own, 1815–1890*, London, Pimlico.

Kynaston, D. (2001) *The City of London*. Vol. IV: *A Club No More, 1945–2001*, London, Chatto & Windus.

Ladipo, D. (1995) Industrial change and social mobility: black men in New York City and London 1970–1990, Paper given at ESRC London Seminar Series, LSE, 31 March.

Land, H. and Lewis, J. (1997) *The Emergence of Lone Motherhood as a Problem in Late Twentieth Century Britain*, WSP 134, STICERD, London School of Economics.

Lawrence, J. (2000) Divisions between rich and poor remain just as rigid as in the dark days of Dickens, *Independent*, 22 December.

Lawrence, R.Z. (1984) Sectoral shifts and the size of the middle class, *The Brookings Review*, Fall, 3–11.

Leadbeater, C. (2000) *Living on Thin Air: the New Economy*, Harmondsworth, Penguin.

Lee, R. and Schmit-Marwede, U. (1993) Interurban competition? Financial centres and the geography of financial production, *International Journal of Urban and Regional Research*, 17, 4, 492–515.

Lee, T.R. (1973) Ethnic and social class factors in residential segregation, *Environment and Planning A*, 477–90.

Lees, L. (1994) Gentrification in London and New York: an Atlantic gap?, *Housing Studies*, 9, 199–217.

Lees, L. and Carpenter, J. (1995) Gentrification in New York, London, Paris: international comparison, *International Journal of Urban and Regional Research*, 19, 2, 286–304.

Levy, P. (1987) The middle class: is it really vanishing?, *The Brookings Review*, Summer, 77–122.

Ley, D. (1980) Liberal ideology and the post-industrial city, *Annals of the Association of American Geographers*, 70, 238–58.

Ley, D. (1981) Inner-city revitalization in Canada: a Vancouver case study, *Canadian Geographer*, XXV, 2, 124–48.

Ley, D. (1996) *The New Middle Class and the Remaking of the Central City*, Oxford, Oxford University Press.

Liebman, R. (2001) East End? Now that's what I call posh, *Independent*, 17 November.

Livingstone, K. (2001a) London needs some more clusters of tall buildings, *Independent*, 13 June.

Livingstone, K. (2001b) The only way is up, *Independent*, 2 November.

Llewelyn-Davies, UCL Bartlett School of Planning (1996) *Four World Cities: a Comparative Study of London, Paris, New York and Tokyo*, for the Department of the Environment and Government Office for London, London, Llewelyn-Davies.

Local Economy Policy Unit (1992) *London and World Cities: Labour Market, Financial Services and Manufacturing*, London and World Cities, LPAC, paper 1.4.

Lockwood, D. (1958) *The Blackcoated Worker*, London, George Allen & Unwin.

Logan, J. (2000) Still a global city: the racial and ethnic segmentation of New York, in P. Marcuse and R. van Kempen (eds) *Globalizing Cities: a New Spatial Order*, Oxford, Blackwell.

Logan, J., Taylor-Gooby, P. and Reuter, M. (1992) Poverty and income inequality, in S. Fainstein, M. Harloe and I. Gordon (eds) *Divided Cities: New York and London in the Contemporary World*, Oxford, Blackwell.

Logan, J., Alba, R. and McNulty, T. (1996) Minorities in global cities: New York and Los Angeles, in E. Mingione (ed.) *Urban Poverty and the Underclass: a Reader*, Oxford, Blackwell.

Lomas, G. (1978) Inner London's future: studies and policies, *The London Journal*, 4, 1, 95–105.

London Business School, the City Research Project (1995) *The Competitive Position of London's Financial Services: Final Report*, Corporation of London.

London County Council (1937) *London's Housing*, London, LCC.

London Development Partnership (1999) *Building London's Economy: a Strategy for the Mayor and the London Development Agency*, London.

London Docklands Development Corporation (1982) *Corporate Plan*, LDDC, London.

London Pride Partnership (1997) *Affordable Homes for London: a Report from the London Pride Partnership*, London, London Pride Partnership.

London Property Research (1996) *Offices to Other Uses*, London Planning Advisory Committee.

London Research Centre (1986) *The Docklands Housing Needs Survey, 1985*, London, LRC.

London Research Centre (1988a) *Council Tenants in London: a Report Based on the Results of the London Housing Survey 1986–87*, London, LRC.

London Research Centre (1988b) *Access to Housing in London: a Report Based on the Results of the London Housing Survey, 1986–87*, London, LRC.

London Research Centre (1989) *The London Housing Survey, 1986–7: Full Report of Results*, London, LRC.

London Research Centre (1992) *Houses into Flats: a Study of Private Sector Conversions in London*, Department of the Environment, London, HMSO.

London Research Centre (1994) *London's Ethnic Minorities: One City, Many Communities*, London, LRC.

London Research Centre (1996) *The Capital Divided: Mapping Poverty and Social Exclusion in London*, London, LRC.

London Research Centre (1997) *London's Workers*, London, LRC.

Lupton, R. (2001) *Places Apart? The Initial Report of CASE's Areas Study*, Centre for the Study of Social Analysis, London, London School of Economics.

Lyons, D. (1994) Changing patterns of corporate headquarter influence, 1974–89, *Environment and Planning A*, 26, 733–47.

Lyons, D. and Salmon, S. (1995) World cities, multi-national corporations, and urban hierarchy: the case of the United States, in P. Knox and P. Taylor (eds) *World Cities in a World System*, Cambridge, Cambridge University Press.

Lyons, M. (1996) Employment, feminisation and gentrification in London, 1981–93, *Environment and Planning A*, 28, 341–56.

Lyons, M. (1999) Professionalisation, feminisation and socio-spatial polarisation in London, 1971–91, *Environment and Planning A*, 31, 493–506.

Lyons, M. and Simister, J. (2000) From rags to riches? Migration and intergenerational change in London's housing market, 1971–91, *Area*, 32, 3, 271–86.

Machimura, T. (1992) The urban restructuring process in Tokyo in the 1980s transforming Toyko into a world city, *International Journal of Urban and Regional Research*, 16, 1, 114–28.

MacKinnon, I. (1991) Housing shortage 'making London a third-rate capital', *Independent*, 20 August.

Major, T. (2001) Deutsche Bank faces up to its identity crisis, *Financial Times*, 28 September.

Marcuse, P. (1986) Abandonment, gentrification and displacement: the linkages in New York City, in N. Smith and P. Williams (eds) *Gentrification of the City*, London, HarperCollins.

Marcuse, P. (1989) Dual city: a muddy metaphor for a quartered city, *International Journal of Urban and Regional Research*, 13, 697–708.

Marcuse, P. (1993) What's so new about divided cities?, *International Journal of Urban and Regional Research*, 17, 355–65.

Markusen, A. and Gwiasda, V. (1994) Multipolarity and the layering of functions in world cities: New York's struggle to stay on top, *International Journal of Urban and Regional Research*, 18, 2, 167–93.

Marriott, O. (1967) *The Property Boom*, London, Hamish Hamilton.

Marsh, D.C. (1965) *The Changing Social Structure of England and Wales, 1871–1961*, London, Routledge & Kegan Paul.

Marsh, G. (1999) *Residential Development in Central and Inner London in 1999*, London, London Residential Research.

Marshall, G. and Rose, D. (1985) Proletarianisation in the British class structure?, *British Journal of Sociology*, 24, 3, 377–96.

Marshall, G., Newby, H., Rose, D. and Vogler, C. (1988) *Social Class in Modern Britain*, London, Hutchinson.

Marshall, J.L. (1968) The pattern of housebuilding in the interwar period in England and Wales, *Scottish Journal of Political Economy*, 184–205.

Marshall, J.N. (1993) Business reorganization and the development of corporate services in metropolitan areas, *Geographical Journal*, 160, 1, 41–9.

Marshall, J.N. and Raybould, S. (1993) New corporate structures and the evolving geography of white collar work, *TESG*, 84, 362–77.

Martin, J.E. (1964) The elements in the industrial geography of Greater London, in J.T. Coppock and H. Prince (eds) *Greater London*, London, Faber.

Martin, J.E. (1966) *Greater London: an Industrial Geography*, London, Bell.

Martin, R. (1988) Industrial capitalism in transition: the contemporary reordering of the British space economy, in D. Massey and J. Allen (eds) *Uneven Redevelopment*, London, Hodder.

Martin, R. (1994) Stateless monies, global financial integration and national economic autonomy: the end of Geography?, in S. Corbridge, R. Martin and N. Thrift (eds) *Money, Power and Space*, Oxford, Blackwell.

Martin, R. (1995) Income and poverty inequalities across regional Britain: the North–South divide lingers on, in C. Philo (ed.) *Off the Map: the Social Geography of Poverty in the UK*, London, CPAG.

Martin, R. and Rowthorn, B. (1986) *The Geography of De-industrialization*, London, Macmillan.

Massey, D. (1988) *Spatial Divisions of Labour*, London, Macmillan.

Massey, D. and Meegan, R. (1978) Industrial restructuring versus the cities, *Urban Studies*, 15, 273–88.

Massey, D. and Meegan, R. (1982) *The Anatomy of Job Loss: the How, Why and Where of Employment Decline*, London, Methuen.

Mathiason, N. (2001) City set for huge expansion, *Observer*, 3 June.

May, J. (1996) Globalization and the politics of place: place and identity in an inner London neighbourhood, *Transactions of the Institute of British Geographers*, 21, 1, 194–215.

Mayhew, H. (1850) *The Morning Chronicle Survey of London*, London.

McCarthy, M. and Thornton, P. (2002) London is now fastest-growing city in Europe. But can it cope?, *Independent*, 12 March.

McDowell, L. (1997) The new service class: housing, consumption and lifestyle among London bankers in the 1990s, *Environment and Planning A*, 29, 2061–78.

Merrell, C. (2000) JP Morgan cuts to the Chase in drive to keep up with rivals, *The Times*, 14 September.

Merrell, C. (2001a) One in five City jobs could be lost, *The Times*, 17 March.

Merrell, C. (2001b) Sony to transfer its treasury operations to Square Mile, *The Times*, June.

Merrett, S. (1979) *State Housing in Britain*, London, Routledge & Kegan Paul.

Merrett, S. (1988) *The Denarius Hypothesis*, Bartlett School of Architecture and Planning, Greater London Housing Study, Working Paper.

Merrick, J. (2001) Size isn't everything, *Independent*, 2 November.

Merrifield, A. (1993) The Canary Wharf debacle: from 'TINA' – there is no alternative – to 'THEMBA' – there must be an alternative, *Environment and Planning A*, 25, 1247–65.

Merrifield, A. and Swynedouw, E. (1996) *Social Justice and the City*, London, Lawrence & Wishart.

Mesure, S. (2001) City strikes back at Canary Wharf, *Independent*, 28 April.

Metcalf, D. and Richardson, R. (1980) Unemployment in London, in A. Evans and D.E.C. Eversley (eds) *The Inner City: Employment and Industry*, London, Heinemann.

Milmo, C. (2000) Fear and loathing in inner city's dark alleyways, *Independent*, 30 November.

Milner Holland (1965) *Report of the Committee on Housing in Greater London*, Cmnd 2605, London, Ministry of Housing and Local Government, HMSO.

Mingione, E. (ed.) (1996) *Urban Poverty and the Underclass: a Reader*, Oxford, Blackwell.

Mitchie, D. (1997) London and the process of economic growth since 1750, *The London Journal*, 22, 63–90.

Mitchie, R.C. (1992) *The City of London: Continuity and Change, 1850–1990*, London, Macmillan.

Modood, T. (1997) *Ethnic Minorities in Britain: Diversity and Disadvantage*, London, Policy Studies Institute.

Mohan, J. (ed.) (1989) *The Political Geography of Contemporary Britain*, London, Macmillan.

Mollenkopf, J. (1998) Changing patterns of inequality in New York City, City University of New York Graduate School, Unpublished paper.

Mollenkopf, J. and Castells, M. (1991) *Dual City: Restructuring New York*, New York, Russell Sage Foundation.

Moore, J. (2000) Property boom sees Savills staff hit jackpot, *The Times*, 5 July.

Moore, J. (2001) Liffe falls to Euronext for £555 million cash, *The Times*, 30 October.

Moore, J. (2002) LSE strikes alliance with Hong Kong, *The Times*, 24 May.

Morrey, C. (1973) *The Changing Population of the London Boroughs*, GLC Department of Planning and Transportation Intelligence Unit, Research Memorandum No. 413.

Morrison, R. (2001) High anxiety: is London's skyline ready for a makeover?, *The Times*, 28 November.

Mortimore, P., Owen, C. and Phoenix, A. (1997) Higher education qualifications, in V. Karn (ed.) *Ethnicity in the 1991 Census*, Vol. IV, London, HMSO.

Mortishead, C. (1999) Airborne Canary leaves ruins of the tower in its wake, *The Times*, 2 March.

Mortishead, C. (2000a) Fast-track Canary aims to snare top tenants, *The Times*, 7 July.

Mortishead, C. (2000b) Enron opts for Canary Wharf headquarters, *The Times*, 25 July.

Moss, M. (1987) Telecommunications, world cities and urban policy, *Urban Studies*, 24, 4, 534–46.

Mullins, A. and Watson-Smyth, K. (2000) It's time you turned us into a hot property, Islington tells the makers of Monopoly, *Independent*, 29 April.

Mullins, P. (1982) The 'middle class' and the inner city, *Journal of Australian Political Economy*, 11, 1, 58.

Mullins, A. and Watson-Smyth, K. (2000) It's time you turned us into a hot property, Islington tells the makers of Monopoly, *Independent*, 29 April.

Munt, I. (1985) Economic restructuring, culture and gentrification: a case study of Battersea, London, *Environment and Planning A*, 19, 1175–97.

Murie, A. and Musterd, S. (1996) Social segregation, housing tenure and social change in Dutch cities in the late 1980s, *Urban Studies*, 33, 3, 495–516.

Murphy, P. and Watson, S. (1994) Social polarization and Australian cities, *International Journal of Urban and Regional Research*, 18, 4, 573–90.

Murphy, R.T. (2000) Japan's economic crisis, *New Left Review*, 1, Jan./Feb., 25–53.

Murray, C. (1984) *Losing Ground: American Social Policy, 1950–1980*, New York, Basic Books.

Murray, C. (1990) *The Emerging Underclass*, London, Institute of Economic Affairs.

Murray, R. and Bianchini, F. (1997) *Policies to Support the Cultural Industries in London*, London Arts Board.

Musterd, S. and Ostendorf, W. (eds) (1998) *Urban Segregation and the Welfare State: Inequality and Exclusion in Western Cities*, London, Routledge.

Myles, J. (1988) The expanding middle class: Canadian evidence on the de-skilling debate, *Canadian Review of Sociology and Anthropology*, 25, 247–67.

Newman, P. and Smith, I. (2000) Cultural production, place and politics on the South Bank of the Thames, *International Journal of Urban and Regional Research*, 24, 1, 9–24.

Noyelle, T. (1983) The rise of advanced services: some implications for economic development in U.S. cities, *American Planning Association Journal*, Summer, 280–90.

Noyelle, T. and Stanbach, T.M. (1985) *The Economic Transformation of American Cities*, Totowa, N.J., Rowman & Allanheld.

O'Brien, R. (1991) *Global Financial Integration: the End of Geography*, Chatham House, London, Pinter.

Office for National Statistics (ONS) (1995) New Earnings Survey, London, HMSO.

Office for National Statistics (2001) *Family Expenditure Survey 2000*, London, HMSO.

O'Leary, D. (1987) Why was the GLC abolished?, *International Journal of Urban and Regional Research*, 11, 193–217.

Olsen, D.J. (1979) *The Growth of Victorian London*, Harmondsworth, Penguin.

Orr, D. (2001) Who benefits from urban regeneration?, *Independent*, 22 June.

Ostendorf, W., Musterd, S. and De Vos, S. (2001) Social mix and neighbourhood effect. Policy ambitions and empirical evidence, *Housing Studies*, 16, 3, 371–80.

Owen, D. and Green, A. (1992) Labour market experience and occupational change amongst ethnic groups in Great Britain, *New Community*, 19, 1, 7–29.

Owen, G. (2002) Parents go private as order collapses in state schools, *The Times*, 8 June.

Pacione, M. (1997) *Britain's Cities: Geographies of Division in Urban Britain*, London, Routledge.

Padgham, J. (2001) Britain's leading bosses collect pay rises of 18 pc, *Evening Standard*, 31 October.

Pahl, R. (1971) Poverty and the urban system, in G. Chisholm and G. Manners (eds) *Spatial Policy Problems of the British Economy*, Cambridge, Cambridge University Press.

Pahl, R. (1988) Some remarks on informal work, social polarisation and the social structure, *International Journal of Urban and Regional Research*, 12, 247–67.

Parker, J. and Dugmore, K. (1976) *Colour and the Allocation of GLC Housing, The Report of the GLC Lettings Survey, 1974–5*, GLC Research Report, 21, Greater London Council.

Parker, J. and Dugmore, K. (1978) Race and the allocation of GLC housing – a GLC survey, *New Community*, 6, 27–41.

Peach, C. (1996) Does Britain have ghettos?, *Transactions of the Institute of British Geographers*, 21, 216–35.

Peach, C. (1997) Pluralist and assimilationist models of ethnic settlement in London, 1991, *Tijdschrift voor Economische en Sociale Geografie*, 88, 120–34.

Peach, C. (1998) South Asian and Caribbean ethnic minority housing choice in Britain, *Urban Studies*, 35, 10, 1657–80.

Peach, C. and Byron, M. (1993) Caribbean tenants in council housing: 'race', class and gender, *New Community*, 19, 3, 407–23.

Peach, C. and Byron, M. (1994) Council house sales, residualisation and Afro-Caribbean tenants, *Journal of Social Policy*, 23, 3, 363–83.

Peach, C. and Shah, S. (1980) The contribution of council housing to West Indian desegregation in London, 1961–71, *Urban Studies*, 17, 333–42.

Pearman, H. (2000) Sky high city dreamers, *Sunday Times*, 3 September.

Peck, J. and Tickell, A. (1992) Local modes of social regulation? Regulation theory, Thatcherism and uneven development, *Geoforum*, 23, 347–63.

Phillips, D. (1986) *What Price Equality? A Report on the Allocation of GLC Housing in Tower Hamlets*, Housing Research and Policy Report 9, London, GLC.

Phillips, D. (1987) Searching for a decent home: ethnic minority progress in the post-war housing market, *New Community*, 14, 1, 105–17.

Phillips, D. (1988) Race and housing in London's East End: continuity and change, *New Community*, 14, 3, 356–69.

Phillips, D. and Karn, V. (1992) Race and housing in a property owning democracy, *New Community*, 18, 3, 349–53.

Pierce, A. (2002) Taxpayer loses huge slice of £700m on Dome giveaway, *The Times*, 30 May.

Pinch, S. (1978) Patterns of local authority housing allocation in Greater London between 1966 and 1973: an inter-borough analysis, *Transactions of the Institute of British Geographers*, NS, 3, 35–54.

Pinch, S. (1993) Social polarisation: a comparison of evidence from Britain and the United States, *Environment and Planning, A*, 25, 779–95.

Power, A. (1993) *Hovels to High Rise: State Housing in Europe since 1850*, London, Routledge.

Power, A. (1996) Area-based poverty and residential empowerment, *Urban Studies*, 33, 9, 1535–64.

Pratt, A. (1997) The cultural industries production system: a case study of employment change in Britain, 1984–91, *Environment and Planning A*, 29, 1953–74.

Preston, V. and McLafferty, S. (1993) Income disparities, employment and occupational changes in New York, *Regional Studies*, 27, 3, 223–35.

Preteceille, E. (1995) Division sociale de l'espace et globalisation: le case de la metropole parisienne, *Sociétés Contemporaines*, 22/23, 33–67.

Preteceille, E. (2001) *Division Sociale et Services Urbains.* Vol. I: *Inégalités et contrastes sociaux en Île de France*, Paris, IRESCO.

Pretzlik, C. and Saigol, L. (2001) Cheerful cull of the richer Masters of the Universe, *Financial Times*, 24 November.

Prince, H. (1964) North-West London 1814–1863, in J.T. Coppock and H.C. Prince (eds) *Greater London*, London, Faber.

Pryke, M. (1991) An international city going 'global': spatial change in the City of London, *Environment and Planning D, Society and Space*, 9, 197–222.

Pryke, M. (1994) Looking back on the space of a boom: (re)developing spatial matrices in the City of London, *Environment and Planning A*, 26, 235–64.

Pryke, M. and Lee, R. (1995) Place your bets: towards an understanding of globalisation, socio-financial engineering and competition within a financial centre, *Urban Studies*, 32, 2, 329–44.

Quah, D. (1997) Increasingly weightless economies, *Bank of England Quarterly Journal*, February.

Raban, J. (1974) *Soft City*, London, Hamish Hamilton.

Raban, J. (1991) *Hunting Mr Heartbreak*, London, Pan Books.

Ramdin, R. (1987) *The Making of the Black Working Class in Britain*, Aldershot, Wildwood House.

Rasmussen, S.E. (1934) *London: the Unique City* (revised edn 1982), London, MIT Press.

Redfern, P. (1997) A new look at gentrification: gentrification and domestic technologies, *Environment and Planning A*, 29, 1275–96.

Reid, I. (1981) *Social Class Differences in Britain* (2nd edn), London, Grant McIntyre.

Reid, T. (2001) Welcome to our mean streets, Mr de Niro, *The Times*, 7 December.

Rhein, C. (1996) Social segmentation and spatial polarisation in Greater Paris, in J. O'Loughlin and J. Friedrichs (eds) *Social Polarisation in Post-Industrial Metropolises*, Berlin and New York, Walter de Gruyter.

Rhein, C. (1998) Globalisation, social change and minorities in Metropolitan Paris, the emergence of new class patterns, *Urban Studies*, 35, 3, 429–48.

Rhodes, J. and Tyler, P. (1998) Evaluating the LDDC: regenerating London's docklands, *Rising East*, 2, 2, 32–41.

Rix, V. (1997) Industrial decline, economic restructuring and social exclusion in London East, 1980s and 1990s, *Rising East*, 1, 1, 118–41.

Roberts, R. (1990) *The Classic Slum: Salford Life in the First Quarter of the Century*, London, Penguin.

Roberts, R. and Kynaston, D. (2001) London will stay a world centre, *Independent*, 10 October.

Robinson, V. (1989) Economic restructuring, the urban crisis and Britain's black population, in D.T. Herbert and D.M. Smith (eds) *Social Problems and the City: New Perspectives*, Oxford, Oxford University Press.

Robson, B. (1969) *Urban Analysis: a Study of City Structure*, Cambridge, Cambridge University Press.

Robson, B. (1975) *Urban Social Areas*, Oxford, Oxford University Press.

Robson, G. and Butler, T. (2001) Coming to terms with London: middle class communities in a global city, *International Journal of Urban and Regional Research*, 25, 1, 70–86.

Rogers, A. (1997) The ethnic division of labour in Greater London: a spatial analysis of workplace–residence relations, in M. Cross and R. Moore (eds) *Globalization and the New City*, Basingstoke, Palgrave.

Rogers, P. (1996) Dresdner chooses London as centre for Eurobond dealing, *Independent*, 16 May.

Rose, D. (1984) Rethinking gentrification: beyond the uneven development of Marxist urban theory, *Environment and Planning D, Society and Space*, 2, 1, 47–74.

Rosser, N. (2000) Madonna comes in from the London cold with a £7m house in Notting Hill, *Evening Standard*, 9 November.

Rubenstein, W. (1977) The Victorian middle classes: wealth, occupation and geography, *Economic History Review*, 30, 100–14.

Rubenstein, W.D. (1986) *Wealth and Inequality in Britain*, London, Faber & Faber.

Rustin, M. (1985) Lessons of the London Industrial Strategy, *New Left Review*, 75–84.

Rutter, M. and Madge, N. (1976) *Cycles of Disadvantage: a Review of Research*, London, Heinemann.

Saigol, L. (2002) West LB set to cut 15% of City jobs, *Financial Times*, 20 May.

Sanai, D. (2000) Clicks and mortar in Clerkenwell, *Independent*, 5 April.

Sarkisson, W. (1976) The idea of social mix in town planning: a historical review, *Urban Studies*, 13, 231–46.

Sarre, P. (1989) Race and the class structure, in C. Hamnett, L. McDowell and P. Sarre (eds) *The Changing Class Structure*, London, Sage.

Sarre, P., Phillips, D. and Skellington, R. (1989) *Ethnic Minority Housing: Explanations and Policies*, Aldershot, Avebury.

Sassen, S. (1986) New York City: economic restructuring and immigration, *Development and Change*, 17, 85–119.

Sassen, S. (1988) *The Mobility of Labour and Capital*, Cambridge, Cambridge University Press.

Sassen, S. (1990) Finance and business services in New York City: international linkages and domestic effects, *International Social Science Journal*, 125, 287–306.

Sassen, S. (1991) *The Global City: New York, London and Tokyo*, Princeton, Princeton University Press (2nd edn 2001).

Sassen, S. (1995) *Cities in a World Economy?*, Thousand Oaks, Calif., Pine Forge Press.

Sassen, S. (1999) Global financial centers, *Foreign Affairs*, 78, 1, 75–87.

Sassen, S. (2000) *The Global City* (2nd edn), Princeton, N.J., Princeton University Press.

Sassen-Koob, S. (1984) The new labour demand in global cities, in M.P. Smith (ed.) *Cities in Transformation*, Vol. 26, Urban Affairs Annual, Beverly Hills, Sage.

Saunders, P. (1979) *Urban Politics: an Introduction*, London, Hutchinson.

Savage, M. (1988) The missing link? The relationship between spatial mobility and social mobility, *The British Journal of Sociology*, 39, 554–77.

Savage, M. and Butler, T. (eds) (1996) *Social Change and the Middle Classes*, London, Taylor & Francis.

Savage, M., Dickens, P. and Fielding, T. (1988) Some social and political implications of the contemporary fragmentation of the 'service class' in Britain, *International Journal of Urban and Regional Research*, 12, 455–76.

Savage, M., Barlow, J., Dickens, P. and Fielding, T. (1992) *Property, Bureaucracy and Culture: Middle Class Formation in Contemporary Britain*, London, Routledge.

Savitch, V. (1988) *Politics and Planning in New York, Paris and London*, Princeton, N.J., Princeton University Press.

Schaffer, R. and Smith, N. (1986) The gentrification of Harlem?, *Annals of the Association of American Geographers*, 76, 3, 347–65.

Scott, A.J. (1997) The cultural economy of cities, *Journal of Urban and Regional Research*, 21, 323–39.

Scott, A.J. (2000) *The Cultural Economy of Cities*, London, Sage.

Scott, A.J. (2001) Capitalism, cities, and the production of symbolic forms, *Transactions of the Institute of British Geographers*, 26, 1, 11–24.

Scott, A., Pearce, D. and Goldblatt, P. (2001) The sizes and characteristics of the minority ethnic populations of Great Britain – latest estimates, *Population Trends*, 105, 6–15.

Seargent, G. (2001) Taxpayers cannot afford skyscrapers, *The Times*, 14 December.

SEJPT (1970) *Strategic Plan for the South East*, Special Consultative Report of the South East Joint Planning Team, London, HMSO.

Sengupta, K. (2000) A terrible place to die. A terrible place to live, *Independent*, 30 November.

Sennett, R. and Cobb, I. (1972) *The Hidden Injuries of Class*, New York, Vintage.

Shah, S. (2000a) No last-minute wobbles as Foster gets permission of 'erotic gherkin', *Independent*, 5 July.

Shah, S. (2000b) Canary Wharf breaks into the black as tenants line up and shares hit record, *Independent*, 21 September.

Shah, S. (2001a) Berkeley unveils £100m plan for UK's tallest block of flats, *Independent*, 13 December.

Shah, S. (2001b) Barclays to quit City for Canary Wharf, *Independent*, 6 September.

Shah, S. (2001c) Minerva plans tower in Square Mile to rival Canary Wharf, *Independent*, 11 September.

Shaw, D. (2001) The post-industrial city, in R. Paddison (ed.) *Handbook of Urban Studies*, London, Sage.

Short, J. (1989) Yuppies, yuffies and the new urban order, *Transactions of the Institute of British Geographers*, 14, 2, 173–88.

Short, J. and Kim, Y. (1999) *Globalization and the City*, Harlow, Addison-Wesley Longman.

Short, J., Kim, Y., Kuss, M. and Wells, H. (1996) The dirty little secret of world cities research: data problems in comparative analysis, *International Journal of Urban and Regional Research*, 20, 4, 697–717.

Silver, H. (1993) National conceptions of the new urban poverty: social structural change in Britain, France and the United States, *International Journal of Urban and Regional Research*, 17, 336–54.

Silverman, G. (2001) Wall Street chiefs receive $30m-plus pay award, *Financial Times*, 19 February.

Simmie, J. (1983) Beyond the industrial city, *Journal of American Planning Association*, Winter, 59–76.

Simmie, J. (1985) The spatial division of labour in London, *International Journal of Urban and Regional Research*, 9, 4, 556–68.

Simmie, J. (ed.) (1994) *Planning London*, London, UCL Press.

Simon, N.W.H. (1977) The relative level and changes in earnings in London and Great Britain, *Regional Studies*, 2, 87–98.

Simpson, S. and Dorling, D. (1994) Those missing millions: implications for social statistics of non-response to the 1991 census, *Journal of Social Policy*, 23, 4, 543–67.

Smith, A. (1989) Gentrification and the spatial contribution of the state: the restructuring of Docklands, *Antipode*, 21, 232–60.

Smith, D.A. and Timberlake, M. (1995) Conceptualising and mapping the structure of the World System's City System, *Urban Studies*, 32, 2, 287–302.

Smith, N. (1979) Toward a theory of gentrification: a back to the city movement by capital not people, *Journal of the American Planning Association*, 45, 4, 538–48.

Smith, N. (1987) Of yuppies and housing: gentrification, social restructuring and the urban dream, *Environment and Planning D, Society and Space*, 5, 151–72.

Smith, N. (1996) *The New Urban Frontier: Gentrification and the Revanchist City*, London Routledge.

Smith, N. (2001) A place for class?, *Environment and Planning A*, 32, 6, 1011–32.

Smith, N. and deFilippis, J. (1999) The reassertion of economics: 1990s gentrification in the Lower East Side, *International Journal of Urban and Regional Research*, 23, 4, 638–53.

Smith, N. and Williams, P. (eds) (1986) *Gentrification of the City*, London, HarperCollins.

Smith, S.J. (1989) *The Politics of 'Race' and Residence: Citizenship, Segregation and White Supremacy in Britain*, Cambridge, Polity Press.

Smitt, R. (1995) EMU the key to City's future, *Independent*, 27 November.

Social Exclusion Unit (SEU) (1998) *Bringing Britain Together: a National Strategy for Neighbourhood Renewal*, Cmnd 4045, London, HMSO.

Soja, E.W. (2000) *Post Metropolis: Critical Studies of Cities and Regions*, Oxford, Blackwell.

Soja, E., Morales, R. and Wolff, G. (1982) Urban restructuring: analysis of social and spatial change in Los Angeles, *Economic Geography*, 59, 2, 195–230.

Sparkes, J. (1999) *Schools, Education and Social Exclusion*, CASE Paper 29, Centre for the Study of Social Exclusion, London School of Economics.

Stark, T. (1992) *Income and Wealth in the 1980s* (3rd edn), Working Group Papers, London, Fabian Society.

Stedman Jones, G. (1971) *Outcast London*, Oxford, Oxford University Press.

Steiner, S. (2000) £200,000 record for average London house, *The Times*, 30 August.

Stevenson, D. (1998) Setting the scene: assessing the impact of the LDDC, *Rising East*, 2, 2, 19–31.

Storkey, M. and Lewis, R. (1996) London: a true cosmopolis, in *Ethnicity in the 1991 Census*, Vol. 3: *Social Geography and Ethnicity in Britain: Geographical Spread, Spatial Concentration and Internal Migration*, ed. P. Radcliffe, London, HMSO.

Sudjic, D. (2001) A thoroughly modernising mayor, *Observer*, 8 July.

Summerson, J. (1978) *Georgian London*, Harmondsworth, Penguin.

Syrett, S. and Baldock, R. (eds) (2001) *Governing London: Competitiveness and Regeneration for a Global City*, London, Middlesex University Press.

Taylor, P. (2000) World cities and territorial status under conditions of contemporary globalization, *Political Geography*, 19, 5–32.

Taylor, P. and Walker, D.R.F. (2001) World cities: a first multivariate analysis of their service complexes, *Urban Studies*, 38, 1, 23–48.

Teedon, P. (2001) Designing a place called Bankside: on defining an unknown space in London, *European Planning Studies*, 9, 4, 459–82.

Thomas, S. and Mortimore, P. (1996) Comparison of value added models for secondary school effectiveness, *Research Papers in Education*, 11, 1, 279–95.

Thompson, G. (1990) *The Political Economy of the New Right*, London, Pinter.

Thornbury, R. (1978) *The Changing Urban School*, London, Methuen.

Thornley, A. (ed.) (1992) *The Crisis of London*, London, Routledge.

Thornley, A. (2000) Dome alone: London's millennium project and the strategic planning deficit, *International Journal of Urban and Regional Research*, 24, 3, 689–99.

Thornton, P. (2000a) How City's high flyers are fuelling the London housing boom, *Independent*, 28 November.

Thornton, P. (2000b) First-time buyers priced out of the London market, *Independent*, 28 December.

Thornton, P. (2001) Income of £30,000 a year now needed to buy a home, *Independent*, 22 August.

Thrift, N. (1987) The fixers: the urban geography of international commercial capital, in M. Castells and J. Henderson (eds) *Global Restructuring and Territorial Development*, London, Sage.

Thrift, N. (1994) On the social and cultural determinants of international financial centres: the case of the City of London, in S. Corbridge, N. Thrift and R. Martin (eds) *Money, Power and Space*, Oxford, Blackwell.

Thrift, N. and Leyshon, A. (1992) In the wake of money: the City of London and the accumulation of value, in L. Budd and S. Whimpster (eds) *Global Finance and Urban Living*, London, Routledge.

Thrift, N. and Leyshon, A. (1994) A phantom state? The detraditionalization of money, the international financial system and international financial centres, *Political Geography*, 12, 299–327. (Reprinted in N. Thrift (1996) *Spatial Formations*, London, Sage.)

Tindall, G. (1973) A street in London, *New Society*, 14 January.

Touche Ross (1994) Statement by Mr John Magill, The Appointed Auditor for Westminster City Council, 19 January.

Toulouse, C. (1992) Thatcherism, class politics and urban development in London, *Critical Sociology*, 18, 1, 55–76.

Townsend, P., Corrigan, P. and Kowarzik, U. (1987) *Poverty and Labour in London: Interim Reports of a Centenary Survey*, London, Low Pay Unit.

Travers, T., Whitehead, C., Holmans, A. and Gordon, I. (2000) *Housing in London: Future Perspectives*, Discussion Paper No. 4, London, London School of Economics.

Turner, D. (2001) London's house prices will be affected by fewer big spenders, *Financial Times*, 24 November.

Ungoed-Thomas, J. (2000) Merger mania bringing City workers a bonus bonanza, *Sunday Times*, 17 December.

Usher, D. (1987) *Housing Privatisation: the Sale of Council Estates*, Working Paper 67, School for Advanced Urban Studies, University of Bristol.

Vallely, P. (1999) They say it's Europe's richest inner city, but try telling them that in Hackney, *Independent*, 11 February.

Varsanyi, M.W. (2000) Global cities from the ground up: a response to Peter Taylor, *Political Geography*, 19, 1, 33–8.

Waldinger, R. (1996) *Still the Promised City? New Immigrants and African Americans in Post-Industrial New York*, Cambridge, Mass., Harvard University Press.

Walker, R. and Greenberg, D. (1982) Postindustrialism and reform in the city: a critique, *Antipode*, 14, 17–32.

Walks, A. (2001) The social ecology of the post-Fordist city? Economic restructuring and socio-spatial polarisation in the Toronto urban region, *Urban Studies*, 38, 3, 407–48.

Waller, M. (2000) City skyline set for more towers after 'gherkin' approval, *The Times*, 24 July.

Walsh, J. (2001) Is this Britain's most des res? Offers invited around £85m, *Independent*, 21 July.

Waples, J. (2000) Giant Exhibition Centre boosts Docklands, *Sunday Times*, 19 November.

Waples, J. (2002) Skyscraper future: City poised for key decision, *Sunday Times*, 9 June.

Ward, D. (1989) *Poverty, Ethnicity, and the American City, 1840–1925: Changing Conceptions of the Slum and the Ghetto*, Cambridge, Cambridge University Press.

Warde, A. (1991) Gentrification as consumption: issues of class and gender, *Environment and Planning D, Society and Space*, 9, 223–32.

Warf, B. (1989) Telecommunications and the globalisation of financial services, *Professional Geographer*, 41, 3, 257–71.

Warf, B. (2000) New York: the Big Apple in the 1990s, *Geoforum*, 31, 4, 487–99.

Warner, J. (2001a) Double standards in Dobson's £10.5 million package, *Independent*, 16 November.

Warner, J. (2001b) The City: British success or a study in failure, *Independent*, 27 October.

Watt, P. (2001) Local authority tenants in Camden, unpublished Ph.D. thesis, Department of Geography, King's College, London.

Watt, P. and Jacobs, K. (2000) Discourses of social exclusion: an analysis of Bringing Britain Together: a national strategy for neighbourhood renewal, *Housing, Theory and Society*, 17, 14–26.

Waugh, P. (2001) Porter is ordered to repay £26.5m in votes scandal, *Independent*, 14 December.

Wessel, T. (2000) Social polarization and socio-economic segregation in a welfare state: the case of Oslo, *Urban Studies*, 37, 11, 1947–68.

Wessel, T. (2001) Losing control: inequality and social divisions in Oslo, *European Planning Studies*, 9, 7, 889–906.

West, A., Power, S., West, R. and Travers, T. (1999) *The Financing of School Based Education – End of Award Report to the ESRC: Main Findings*, London, Centre for Educational Research.

Westaway, J. (1974) The spatial hierarchy of business organisation and its implications for the British urban system, *Regional Studies*, 8, 145–55.

Western, J. (1993) Ambivalent attachments to place in London: twelve Barbadian families, *Environment and Planning D, Society and Space*, 11, 147–70.

Wheatcroft, P. (2000) Canary Wharf: Success from near disaster, *The Times*, 21 September.

White, J. (1986) *The Worst Street in North London: Cambell Bunk, Islington, Between the Wars*, London, Routledge & Kegan Paul.

White, P. (1993) The social geography of immigrants in European cities: the geography of arrival, in R. King (ed.) *The New Geography of European Migration*, London, Belhaven.

White, P. (1998) The settlement patterns of developed world migrants in London, *Urban Studies*, 35, 10, 1725–44.

Whitehand, J. (1988) The changing urban landscape: the case of London's high-class residential fringe, *The Geographical Journal*, 154, 3, 351–66.

Whitehead, M. (1992) The health divide, in *Inequalities in Health*, Harmondsworth, Penguin.

Wilcock, J. and Eisenhammer, J. (1994) Deutsche to integrate in London, *Independent*, 9 November.

Williams, C. and Windebank, J. (1995) Social polarization of households in contemporary Britain, *Regional Studies*, 29, 8, 723–8.

Williams, P. (1976) The role of institutions in the inner London housing market: the case of Islington, *Transactions of the Institute of British Geographers*, NS, 1, 1, 72–82.

Williams, P. (1978) Building societies and the inner city, *Transactions of the Institute of British Geographers*, NS, 3, 23–34.

Williams, S. (1992) The coming of the groundscapers, in L. Budd and S. Whimpster (eds) *Global Finance and Urban Living*, London, Routledge.

Wilmott, P. (1969) Some social trends, *Urban Studies*, 6, 286–308.

Wilmott, P. and Young, M. (1973) Social class and geography, in D. Donnison and D. Eversley (eds) *London: Urban Patterns, Problems and Policies*, London, Heinemann.

Wilson, W.J. (1987) *The Truly Disadvantaged: the Inner City, the Underclass and Public Policy*, Chicago, University of Chicago Press.

Wilson, W.J. (1996) *When Work Disappears: the World of the New Urban Poor*, New York, Vintage Books.

Wohl, A. (1971) The housing of the working classes in London, 1815–1914, in S.D. Chapman (ed.) *A History of Working Class Housing*, Newton Abbot, David & Charles.

Wolf, T. (1987) *The Bonfire of the Vanities*, New York, Farrar, Strauss & Giroux.

Wood, C. (1992) *The Bubble Economy: the Japanese Economic Collapse*, London, Sidgwick and Jackson.

Wood, S. (ed.) (1982) *The Degradation of Work: Skill, Deskilling and the Labour Process*, London, Hutchinson.

Woods, R. and Godson, R. (2001) City bankers face pay cuts of 60%, *Sunday Times*, 11 November.

Woolf, M. and Boggan, S. (2002) Dome goes for nothing in exchange for profits, *Independent*, 30 May.

Wright, E. and Martin, B. (1987) The transformation of the American class structure, 1960–1980, *American Journal of Sociology*, 93, 1–29.

Wright E.O. and Singlemann, J. (1982) Proletarianisation in the changing American class structure, *American Journal of Sociology*, 88 (special supplement), 176–289.

Wright, P. (1985) *On Living in an Old Country: the National Past in Contemporary Britain*, London, Verso.

Wright, P. (1991) *A Journey Through Ruins: the Last Days of London*, London, Paladin.

Wrigley, E.A. (1967) A simple model of London's importance in changing English society and economy, *Past and Present*, xxxviii, 44–70.

Wyly, E.K. and Hammel, D.J. (1999) Islands of decay in seas of renewal: housing policy and the resurgence of gentrification, *Housing Policy Debate*, 10, 4, 711–71.

Young, K. and Garside, P.L. (1982) *Metropolitan London: Politics and Urban Change, 1837–1981*, London, Heinemann.

Young, K. and Kramer, J. (1978) *Strategy and Conflict in Metropolitan Housing*, London, Heinemann.

Zukin, S. (1987) Gentrification, culture and capital in the urban core, *Annual Review of Sociology*, 13, 129–47.

Zukin, S. (1988) *Loft Living: Culture and Capital in Urban Change*, London, Radius.

Zukin, S. (1992) The City as a landscape of power: London and New York as global financial capitals, in L. Budd and S. Whimpster (eds) *Global Finance and Urban Living*, London, Routledge.

Zukin, S. (1995) *The Culture of Cities*, Oxford, Blackwell.

Index

Page references for figures, tables and plates are in *italics*.

New York 7, 13, 22, 23, 24, 51;
development 213; digital economy
31; ethnic minorities 121, 127;
financial and business services 27, 37;
gentrification 178, 179–80; housing
11; inequality 74, 99, 102, 169;
labour market 10; loft living 144,
222; migration 126; occupational
class structure 70; social polarisation
59, 60, 61; stock exchange *37*, 38,
46
Newham: ethnic minorities 112, 124;
house prices 154; housing 138; Social
Security benefits 195; unemployment
193
Newman, P. 246
Newsweek 2
North Kensington 176
North Southwark Community Action
Group 235
Northern Trust Company 239
Notting Hill 1; ethnic minorities 110,
163; gentrification 2, 161, 162, 166,
185, 247
Notting Hill 2
Noyelle, T. 5, 57, 76

Observer 229
occupational class structure 18, 48–50,
51–2, *64*; changes 7–8, 62–7, *65*, *66*,
67, *68*, 69–71, *69*; ethnic minorities
116–17, *116*, 127; and housing tenure
131, 147–50, *149*, *150*, *151*, 157–8;
and inequality 9; national differences
9–10; professionalisation 55–7;
proletarianisation 52–5; social
polarisation 57–62
office buildings: conversions 144, 145,
219, 221–2; development 223, 225–6,
226, *227*, 228, 245–6; development
policy 33, 228–9, *230*, 231; high-rise
231–2; *see also* Canary Wharf
Office for National Statistics (ONS) 36,
36, *96*
Office of Population, Censuses and
Surveys (OPCS) *64*, 65
Ogden, P. 165, 167
O'Leary, D. 14
Olsen, D.J. 200, 201
Olympia and York 238, 241
one-parent families 118
options trading 37

'Oranges and Lemons' 247
Orr, D. 179
Osman, Kayser 208
Ostendorf, W. 9, 11
Owen, D. 109, 115
owner-occupation *see* home ownership

Paddington Basin 245–6
Pahl, R. 75
Pakistanis 110, 111–12, *111*, *112*;
education 118; employment 115;
household incomes 117, *117*; housing
122, *123*; occupational class structure
116, 117; unemployment 119, *120*
Palumbo, Peter 228
Paris 22, 24, 41, 42; digital economy
31; inequality 74, 99, 102; migration
103; occupational class structure
70
Parker, J. 123, 130
part-time workers 78
Peach, C. 108, 115, 122–3, 124, 126,
130, 185
Pearman, H. 245–6
Pearson Longman 43
Peck, J. 16
Peckham 208, 234–5
Petrus 73
Phillips, D. 123, 130
Pierce, A. 217
planning 13–15, 17, 33, 213, 222
polarisation: incomes 8, 73, 100, 102;
measuring 77–8; *see also* earnings
polarisation, social polarisation
police 209
politics 2, 4, 13–15
Pollard, S. 199
population 4, 103, 156, 161
Porter, Dame Shirley 145–7
post-industrialism 5–7, 18, 21–2;
employment 63, 69; housing
184; landscape 19, 211, 218;
proletarianisation theories 53–4
poverty 189, 191, 196–8, 204–5, 210
power 24
Power, A. 134, 222
Pratt, A. 28, 36, *36*
Preteceille, E. 70, 74, 99
Pretzlik, C. 39, 42, 95
Price Waterhouse Coopers 93
Primrose Hill 161, 162, 172
Prince, H. 132